EXPENDABLE AMERICANS

ALSO BY PAUL BRODEUR

Fiction

The Sick Fox

The Stunt Man

Downstream

Nonfiction

Asbestos and Enzymes

EXPENDABLE AMERICANS

Paul Brodeur

THE VIKING PRESS NEW YORK

First published in a hardbound and a paperbound edition
in 1974 by The Viking Press, Inc.
625 Madison Avenue, New York, N.Y. 10022
Published simultaneously in Canada by
The Macmillan Company of Canada Limited

Library of Congress Cataloging in Publication Data

Brodeur, Paul.
Expendable Americans.

"The material in this book appeared originally as a series of articles in *The New Yorker*, in slightly different form."
1. Industrial hygiene—Addresses, essays, lectures. 2. Industrial toxicology—Addresses, essays, lectures.
I. Title.
RC967.B69 1974 613.6'2'0973 74-4549
ISBN 0-670-30200-7
ISBN 0-670-00580-0 (pbk.)

This book is printed on 100% recycled paper
Printed in U.S.A.

TO THE MEN OF THE TYLER PLANT

ACKNOWLEDGMENTS

The author wishes to express his appreciation to William Shawn, the editor of *The New Yorker*, who provided him with time to follow this project through to the end, and space in the magazine to publish it; to Charles Patrick Crow, of *The New Yorker*, who edited the manuscript; and to Anne Mortimer-Maddox, whose valuable research insured that the manuscript would be as free of error as we could make it.

CONTENTS

PART ONE

Some Nonserious Violations

In the winter of 1972, a flurry of unusual activity accompanied the closing of a factory owned by the Pittsburgh Corning Corporation in Tyler, Texas, a city of sixty thousand about a hundred miles east of Dallas. Production stopped on February 3, 1972, and then the factory, which for more than seventeen years had been manufacturing asbestos insulation, was subjected to a cleanup of prodigious scope and intensity. Under the scrutiny of armed Pinkerton guards, who had been hired by the company to keep unauthorized persons away from the plant and its nearby dumps, sixty-two employees spent a week scraping asbestos waste from machinery and other equipment, and removing and burying truckload after truckload of asbestos scrap that had accumulated in the plant. This work force was laid off permanently on February 11th, and a crew of four maintenance men spent the next two weeks washing down ceilings and walls and steam-cleaning every piece of machinery in sight. Meanwhile, thirty-five thousand burlap sacks, in which amosite asbestos had been shipped to the factory from mines in South Africa, and which, once emptied, had been sold for a nickel apiece to some of the rose nurseries for which the Dallas-Tyler area is famous, were repurchased by Pittsburgh Corning at double the price, brought back to the plant in trucks, and buried at one of the factory's dumps. Toward the end of February, the skeleton crew, using acetylene torches, cut up three one-hundred-and-fifty-foot-long chain conveyor belts, three five-hundred-pound cyclone machines, and hundreds of feet of ventilation pipe, all of which were then taken outside and buried in a dump. Other pieces of heavy equipment, including eight twelve-foot-high feeding machines, three one-hundred-foot-long drying ovens, and a dozen dust collectors, were cut up and sold for junk, and still other items, such as saws, an asbestos-scrap grinder, and the draw works and gears for the ovens, were shipped by rail to Pittsburgh Corning's home office, in Pittsburgh. By the end of March, the Pinkertons were no longer needed—there being very little left of the Tyler plant for any trespasser to see—and by the end of April practically nothing remained of the factory except two dilapidated wooden buildings, which had once been warehouses at Camp Fannin, a World War II training center and POW camp. One of them—the production building—was virtually empty; the other, the factory's storage area, was half filled with sacks of amosite-asbestos fiber, which the company had not used before shutting the plant.

Although Pittsburgh Corning never gave any official reason for the drastic tidying up that accompanied the closing of its Tyler plant, an explanation of the shutdown itself was made two weeks before by E. W. Holman, the corporation's vice-president in charge of manufacturing and technology. In an interview published in the Tyler *Courier-Times* on January 19th, Holman said that increased costs of buying amosite-asbestos fiber and transporting it from South Africa, plus the fact that his company was finding it more and more difficult to market its finished product because of competition, had brought about a decision to cease operations at the Tyler plant and at a Pittsburgh Corning plant in Port Allegany, Pennsylvania. He remarked that "new clean-air restrictions" had also played a role in "speeding up" the closing of the two factories, since the restrictions would have forced his company to install costly filtering equipment in order to bring the level of asbestos dust in the factories down to lawful limits. According to the *Courier-Times*, Holman acknowledged that asbestos fibers posed a health hazard, but added that he knew of no specific Pittsburgh Corning employee who was suffering from significant illness as a result of working with the product.

Three weeks after Holman's remarks were published, a somewhat different version of the situation at the Tyler plant was given by Anthony Mazzocchi, the director of the Legislative Department of the Oil, Chemical, and Atomic Workers International Union, which had represented employees at the factory since 1962. Speaking at a press conference in Washington, D.C., on February 10th, Mazzocchi disclosed that a government survey had shown major industrial-hygiene deficiencies in the operation of the plant, including a grossly inadequate ventilation system, which had resulted in airborne-asbestos levels constituting a critical occupational-health hazard. According to Mazzocchi, the survey had determined that seven of the eighteen workers who had been employed at the factory for ten years or more showed symptoms of asbestosis—scarring of the lungs caused by inhalation of asbestos fibers—which is a significant illness by almost any standard, in that it is irreversible, untreatable, often disabling, and frequently fatal. Pointing out that asbestos has also been proved to be a potent carcinogen, Mazzocchi voiced the fear that many of the men who had worked in the factory would one day be afflicted with lung cancer or other malignant tumors. Morever, he indicated that a health hazard might extend far beyond the plant, because the company had sold tens of thousands of burlap sacks contaminated with asbestos dust to nurserymen, who used them to wrap evergreens and other stock for shipment to retailers and gardeners throughout the nation.

For its part, Pittsburgh Corning made a public response only to the last of Mazzocchi's disclosures. On February 16th, a spokesman for the company admitted to a reporter for the Tyler *Morning Telegraph* that thirty-five thousand burlap bags had been recalled from the Dallas-Tyler area, but he denied that there was any reason to consider them a health hazard. The apparent contradiction in this statement was not resolved by the manager of the Tyler plant, who was quoted at the same time as defying anyone to find any of the bags in question. "We've hired bulldozers to put all those bags underground," he said, making an assertion that would soon apply to much of the factory equipment as well. By then, however, word had got out that considerably more of the Tyler plant was buried than burlap bags and cut-up machinery.

In a sense, the story of the Tyler plant begins with the founding of the Union Asbestos & Rubber Company, in Chicago, in 1918. According to the United States Bureau of Labor Statistics, American and Canadian insurance companies were even then generally declining to insure asbestos workers because of the assumed hazardous conditions of the asbestos industry. Union Asbestos started out as a jobber of railway supplies and an assembler of finished asbestos and rubber products. Business expanded rapidly, thanks to the development of a flexible asbestos tape, which achieved wide use for insulating pipes in steam locomotives, and in 1926 Union Asbestos built a factory in Cicero, Illinois, to manufacture asbestos textiles, insulation materials, packings, brake linings, and gaskets, and a variety of rubber products. Another leap forward took place in the mid-thirties, when the company developed an amosite-asbestos pipe insulation for the Navy.

Amosite is a variety of asbestos found in large deposits in the Transvaal region of South Africa, and it had never been used before in the United States, where most asbestos products had been (and continue to be) made of chrysotile, a variety of the mineral that exists in vast deposits in Canada and the Soviet Union, and accounts for ninety-five per cent of the world's production. Because it is as heat-resistant as chrysotile, and can be purchased more cheaply, amosite was chosen for insulating the pipes, turbines, and boilers of modern warships, and by 1940 the Navy's demands for amosite pipe insulation were such that Union Asbestos—or UNARCO, as it had come to be known—started a plant in Paterson, New Jersey. During the war years, the UNARCO plants in Cicero and Paterson churned out amosite pipe covering for the Navy around the clock, winning numerous Army-Navy "E" awards. Such insulation continued to be much in demand in the postwar period, and in 1949 the company set up a third plant to manufacture it, in McGregor, Texas. Then, in November of 1954, as part of a consolidation program, the company shut the McGregor and Paterson factories, and opened the factory at Tyler, Texas.

Little was known about the Tyler plant except that it was set up to operate generally like the factory in Paterson. However, some information that would one day impart tremendous medical significance to this similarity was just then beginning to be developed by Dr. Irving J. Selikoff, a chest physician, who is head of the Division of Environmental Medicine at the Mount Sinai School of Medicine of the City University of New York, director of its Environmental Sciences Laboratory, and a pioneer in the field of asbestos epidemiology. A native of New York City, Dr. Selikoff interned at the Beth Israel Hospital in Newark; did his pathology work at Mount Sinai, where he has been a member of the staff since 1947; and became a chest physician at the Sea View Hospital, on Staten Island, specializing in tuberculosis. In 1951, he participated in the basic research on isoniazid—the antibiotic drug that, by effectively killing tubercle bacilli, has provided a cure for tuberculosis—and in

1953 he founded a medical clinic in Paterson, where, by chance, seventeen of his early patients were men who worked in the nearby UNARCO plant. At the time, fifteen of the men showed some evidence of pulmonary defects resulting from the inhalation of asbestos. When the Paterson factory closed, they went into other work, and at that point Dr. Selikoff decided to continue his observation of them with X-ray examinations and lung-function tests to determine the history and the natural course of asbestosis in men who would not be further exposed, but in whose tissues the previously inhaled fibers would remain. This was the start of a long journey of discovery for Dr. Selikoff, who would eventually help to demonstrate that asbestos is one of the major industrial causes of cancer. At the time, he was interested chiefly in asbestosis, because he was not convinced that the relationship between asbestos and cancer, which had previously been suggested by a number of medical authorities, would prove to be a serious problem. As things turned out, he changed his mind. In 1954, all seventeen men from the Paterson factory were working and apparently able-bodied. In 1974, only two of them are alive. Of the fifteen who died, seven were victims of lung cancer, two of cancer of the stomach, four of asbestosis, and one of malignant mesothelioma—an invariably fatal tumor of the pleura, the membrane that encases the lung, or of the peritoneum, a similar membrane that lines the abdominal cavity—which rarely occurs without some exposure to asbestos. (One of the fifteen deaths was of heart disease.) As early as 1961, by which time six of the seventeen had died, Dr. Selikoff began to suspect the worst for men who were occupationally exposed to the mineral. At that time, he wrote to Edwin E. Hokin, the president of UNARCO, asking him to make employment records available, so that he could undertake a survey of all the men who had worked in the Paterson factory. Hokin turned the request down, saying the records were not available, but he was surely aware that men who had worked in the Paterson factory might be having medical problems, for during the nineteen-fifties the company had paid out substantial amounts of money to employees of the plant who had become disabled with asbestosis. Dr. Selikoff then wrote to several other large asbestos manufacturers in the United States to ask about the health experience in their plants, and was unable to obtain information from any of them. Meanwhile, he and Dr. Jacob Churg, the chief pathologist at Barnert Memorial Hospital, in Paterson, and a member of the staff of Mount Sinai Hospital, who had himself been finding asbestosis and lung cancer in a large number of workers from the Paterson factory, took their data to Dr. Roscoe P. Kandle, the Commissioner of the New Jersey State Department of Health. Concerned about the situation, Dr. Kandle applied to the United States Public Health Service for funds to undertake a study of the Paterson plant and to make a statewide survey to determine how many people were being occupationally exposed to asbestos. However, the request was denied, lack of funding being given as the reason.

Since Dr. Selikoff already knew that men who had worked in an insulation factory were dying of asbestosis or cancer at an alarming rate, he felt that men who were installing such materials might also risk disease, and early in 1962 he made contact with officials of New York Local 12 and Newark Local 32 of the International Association of Heat and Frost Insulators and Asbestos Workers. The asbestos insulators had been trying for years without success to interest doctors and various government agencies in their medical problems, so they were only too glad to cooperate, and they urged Dr. Selikoff to study the effects of asbestos exposure among their members. He accepted the responsibility, and, though continuing to monitor those of his original seventeen patients who had survived, temporarily abandoned his project to study the other men who had worked in the Paterson factory. At the time, he did not know of the existence of the UNARCO plant in Tyler, Texas.

The Tyler plant was then thriving but was about to change hands. UNARCO held a large contract with the Navy to provide pipe covering for atomic submarines, and the factory was also producing insulation for the chemical-processing industry on the nearby Gulf Coast, which was growing rapidly. Over the years, however, the company had acquired half a dozen plants for the manufacture of various products unrelated to asbestos, and this diversification had altered the objectives of its managers, who decided to quit the asbestos business altogether. As a result, the company sold the Tyler plant in 1962 to the Pittsburgh Corning Corporation—a joint venture of the Pittsburgh Plate Glass Company (now called PPG Industries) and the Corning Glass Works—and the production of amosite-asbestos pipe covering continued as before. By the summer of 1963, however, the new owners were apparently entertaining some misgivings about working conditions in the factory, for at that time they asked the Industrial Hygiene Foundation of America to evaluate the asbestos-dust hazard there. The foundation, which is in Pittsburgh, describes itself as "an association of industries for the advancement of healthful working conditions," and it is financed entirely by industry. It sent industrial-hygiene engineers to the Tyler plant in July and August to review the potential health hazards of handling asbestos and to take samples of airborne asbestos-dust concentrations. In its report to Pittsburgh Corning, the foundation made no mention of any health hazard, and assured the company that, except in a few areas, the number of asbestos fibers found in the air of the Tyler plant was well below the threshold limit value of five million particles per cubic foot of air—a safety standard for dust in asbestos factories that had been adopted in 1946 by the American Conference of Governmental Industrial Hygienists, which, despite its imposing title, is not a government agency but a voluntary organization with members from various groups, including industry, and with the self-imposed task of recommending safety standards for hazardous substances in industry. Incredibly, the authors of the foundation's report appear to have based their judgment on the assumption that the threshold limit value of five million particles per cubic foot meant five million asbestos fibers, whereas the proponents of the threshold limit value had intended it to apply to *all* the particulate matter—fibrous and nonfibrous—in a given cubic foot of air. To understand the magnitude of this error, it should be noted that the study upon which the standard was based had been made in the winter of 1935–36 in four asbestos-textile plants where asbestos fibers were found to constitute about ten per cent of the total amount of airborne dust. The asbestos fibers in the airborne dust measured in the Tyler plant by engineers of the Industrial Hygiene Foundation ranged from a low of twenty-nine per cent to a high of fifty-six per cent. The foundation, however, reported the percentages as if they were of little or no consequence, and contented itself with making recommendations for better housekeeping, better ventilation equipment, and improved maintenance of the ventilation system.

These measures were desperately needed, but there is little evidence to suggest that Pittsburgh Corning felt compelled to initiate them, for when the next survey of the plant was made, more than three years later, conditions were even worse. By that time, the company had acquired a new medical consultant—Dr. Lee B. Grant, a retired colonel, who had been Chief of Aerospace Medicine for the United States Air Force Logistics Command, and who had become medical director of one of Pittsburgh Corning's parent corporations, the Pittsburgh Plate Glass Company, in 1965. At Dr. Grant's request, a survey of the Tyler plant was conducted in November of 1966 by J. T. Destefano, safety and industrial-hygiene engineer for the glass division of Pittsburgh Plate Glass, to see if there had been any significant change in the levels of airborne-asbestos dust since the 1963 survey. After analyzing samples of air from sixteen different areas of the plant, Destefano subsequently reported that asbestos-fiber counts exceeded the threshold limit value in seven instances and that in three of the samples the count was twenty million or more fibers per cubic foot of air. Destefano was, apparently, making the same erroneous assumption about the meaning of the threshold limit value which had been made three years before by engineers of the Industrial Hygiene

Foundation. As a result, though he also suggested better ventilation equipment and improved maintenance of the ventilation system, his report did not mention that workers at the Tyler plant were breathing concentrations of asbestos fibers ten times greater than those of the recommended safety standard that was supposed to protect them from disease.

During the span from 1963 to 1966, a tremendous amount of new information concerning the biological effects of asbestos had been developed and was being circulated through the medical and industrial communities. Perhaps the most important study of the period was the one that Dr. Selikoff conducted of the asbestos insulators. As asbestos workers go, these men had comparatively light and intermittent exposure: they often worked out-of-doors on construction projects; they spent half their time working with materials other than asbestos; and most of the asbestos materials they used had an asbestos content of less than fifteen per cent. (The men at the Tyler plant worked in a confined, far dustier atmosphere, and manufactured a product that had an asbestos content of almost ninety per cent.) In spite of this relatively light exposure, however, Dr. Selikoff found radiological evidence of fibrosis of the lungs—that is, scarring of the lungs—in fully half of eleven hundred and seventeen members of the two locals of the Heat and Frost Insulators and Asbestos Workers. Moreover, among three hundred and ninety-two men with more than twenty years of experience in the trade, he found that three hundred and thirty-nine had developed asbestosis, and that the disease had by then become moderate or extensive in more than fifty per cent of the cases.

Even more alarming were the results of a mortality study of workers in the two locals. During the early part of 1962, Dr. Selikoff and his administrative assistant, Mrs. Janet S. Kaffenburgh, pored over the union records and compiled a list of the names and addresses of all the six hundred and thirty-two men who had been members of the locals on December 31, 1942, and of the eight hundred and ninety men who had joined between then and December 31, 1962. From the union employment records, they obtained detailed work histories of the total membership of fifteen hundred and twenty-two men, including data on the men's leaving work for other employment, war service, illness, or retirement. This enabled them to calculate the onset and duration of exposure to asbestos for each worker. Records of the union health-and-welfare funds provided them with the dates and places of death of two hundred and sixty-two workers who had died between 1942 and 1963, and copies of the death certificates of all but one of them were obtained. In addition, autopsy protocols, histological specimens, and hospital records were reviewed by Dr. Selikoff and Dr. Churg in those deaths (approximately half the total) which occurred in hospitals.

In the next phase of the study, Dr. Selikoff and Dr. Churg were joined by Dr. E. Cuyler Hammond, vice-president for epidemiology and statistics of the American Cancer Society, who had participated in an analysis of the medical effects of the atomic explosions that devastated Hiroshima and Nagasaki in 1945, and whose large-scale epidemiological studies of more than a million men and women provided a major basis for the conclusions drawn in the 1964 Surgeon General's report on the effects of cigarette smoking. Since previous studies had suggested that lung cancer associated with asbestosis seldom develops until twenty years after initial exposure to asbestos dust, Dr. Selikoff and Dr. Hammond decided to limit their first analysis to the six hundred and thirty-two men who were on the union rolls as of December 31, 1942. Taking the men's ages into consideration, Dr. Selikoff and Dr. Hammond then set about comparing the number and causes of death among them with those of the general male population in the United States. The results were depressing. According to the standard mortality tables, two hundred and three deaths could have been expected among the six hundred and thirty-two workers. Instead, there were two hundred and fifty-five, not counting seven men who had died before incurring twenty years of exposure—an excess of twenty-five per cent. The reason for the excess was not hard to find. The fact that twelve of the deaths were attributed to asbestosis was not particularly surprising, but where six or seven deaths from cancer of the lung, pleura, or trachea were to be expected, there were actually forty-five. And where nine or ten gastrointestinal cancers were to be expected, there were twenty-nine. Since the death rate from lung cancer was known to be more than ten times as high among cigarette smokers as among nonsmokers, Dr. Selikoff and Dr. Hammond realized that they would have to take the smoking habits of the asbestos-insulation workers into account if their findings were to have solid validity. It was, of course, impossible for them to ascertain this information with accuracy in the cases of the two hundred and fifty-five men who had died, so, for purposes of calculation, they assumed that all six hundred and thirty-two men had smoked a pack or more of cigarettes each day, and they demonstrated that even if this had been the case it would have produced a lung-cancer death rate only three and a half times that of the general male population. Cigarette smoking, therefore, could not explain the fact that in this group of asbestos-insulation workers the rate of death from lung cancer was seven times the expected rate.

Because of this study's objectivity, its scope, and its thoroughness, it had a great impact on the medical community. The findings were reported to the annual convention of the American Medical Association in June of 1963—a month before the Industrial Hygiene Foundation began its survey of the Tyler plant—and they were published in the spring of 1964 in the *Journal of the American Medical Association*. It was the first study ever made that had taken a large enough group of asbestos workers from a point far enough back in time and followed them long enough to determine unequivocally what their health experience had been. Unlike almost all the previous investigations, which indicated simply that there was a connection between asbestos and various kinds of cancer, it was based upon the incidence of disease within a defined population, and thus answered a fundamental epidemiological question of how many cancers had developed among how many persons exposed. In doing so, it furnished the first incontrovertible evidence that industrial exposure to asbestos was hazardous; it established sound methodology for future studies; and it marked a turning point in the views held by doctors and health officials around the world.

In October of 1964, in order to review the data that had already been collected and to discuss the problems awaiting solution, the New York Academy of Sciences sponsored an international Conference on the Biological Effects of Asbestos, which was held at the Waldorf-Astoria and was attended by more than four hundred scientists. In addition to the statistics provided by Dr. Selikoff and Dr. Hammond on the incidence of asbestosis and cancer in the insulation workers, there were dozens of reports on the occurrence of disease in people exposed to asbestos. Some of the most alarming information was provided by Dr. J. G. Thomson, of South Africa, who reported finding what appeared to be asbestos bodies—inhaled fibers that have been altered by the reaction of lung tissue, and coated with a colloidal substance rich in iron—in the lungs of one in four people coming to autopsy at random in Cape Town. Although asbestos bodies are regularly seen in the lungs of asbestos workers, this discovery indicated that asbestos was becoming a common contaminant in the community at large. There was also a report that mesothelioma was afflicting people who had had only minor exposure to asbestos. This tumor, which takes from twenty to forty years to develop, was previously so rare that it was known to occur in only about one in ten thousand deaths in the general population. By the time of the international conference, however, it was being found increasingly—not only in people who were exposed to asbestos in their work but also in people who lived in the vicinity of asbestos mines and dumps, or factories where asbestos products were manufactured, or who simply lived in the same house with workers who came home with asbestos dust on their clothes. Perhaps the most striking confirmation of this came from London, where Dr. Muriel L.

Newhouse, of the Department of Occupational Health at the London School of Hygiene and Tropical Medicine, investigated seventy-six cases of mesothelioma that had been ascertained by autopsy or biopsy in the London Hospital. To no one's surprise, thirty-one of the seventy-six patients had worked with asbestos, but, in addition, eleven of the forty-five who had not worked with asbestos had simply lived within half a mile of an asbestos factory, and nine others—seven women and two men—were relatives of asbestos workers. Most of these women had washed their husbands' work clothes regularly. Both of the men in this group, when they were boys of eight or nine, had had sisters who worked in asbestos-textile factories. One of the sisters had been employed as a spinner from 1925 to 1936, and had died of asbestosis in 1947, at which time it was determined at an inquest that "she used to return from work with dust on her clothes." Her brother, who had apparently had no other sustained exposure to asbestos in his lifetime, died in 1956 of a pleural mesothelioma. Subsequently, in the United States, there were similar findings in a number of places. For example, the proprietor of a junkyard next to the UNARCO factory in Paterson died of mesothelioma, and so did the engineer who first developed the amosite pipe covering manufactured by the company for the Navy, as did his daughter, whose only known exposure to the mineral was that she sometimes played with samples of asbestos products her father brought home for his family to examine. As a result of such incidents, scientists were forced to revise their idea that asbestos was only an industrial hazard, and to give serious consideration to Dr. Thomson's prediction of danger to untold numbers of people in the general community. Such consideration proved to be well founded, for since then the presence of asbestos bodies in the lungs of ordinary urban dwellers has been confirmed by studies made in Miami, London, Belfast, Pittsburgh, and New York City, where, in a recent investigation conducted by Dr. Arthur M. Langer, the chief mineralogist at the Mount Sinai Environmental Sciences Laboratory, electron-microscope examination of representative samples of tissue showed chrysotile asbestos to be present in the lungs of a hundred and four out of a hundred and twenty-eight people coming to autopsy at random in three city hospitals.

The attitude of the asbestos industry at this time can perhaps be best illustrated by a cautionary letter that was sent to Mrs. Eunice Thomas Miner, the executive director of the New York Academy of Sciences, on October 26, 1964, just after the Conference on the Biological Effects of Asbestos ended, by lawyers representing the Asbestos Textile Institute—an association of asbestos manufacturers that includes the Johns-Manville Corporation, Raybestos-Manhattan, Inc., and Uniroyal, Inc. The letter began by stating that all member companies of the institute shared a grave concern over recent articles carried in local and national newspapers concerning mesothelioma. It went on to say that "innocent but unwise treatment of research data in public discussions, or leaving it to laymen to appreciate the carefully phrased limitations and qualifications, can cause reactions that are not justified by the state of scientific knowledge," and it urged caution in the discussion of medical research into asbestos disease, "to avoid providing the basis for possibly damaging and misleading news stories." It concluded by warning the New York Academy of Sciences that although the right to discuss these subjects was clear, "the gravity of the subject matter and the consequences implicitly involved impose upon any who exercise those rights a very high degree of responsibility for their actions."

During 1966, the academy sent out thousands of copies of its report of the conference to doctors, officials of state and federal health agencies, and custodians of medical libraries all over the country, and from 1965 on there were many articles in leading medical journals and dozens of newspaper stories concerning new and alarming data that had been developed about the perils of inhaling asbestos. As a result, it seems highly probable that by late 1967 the industrial-health officer of any responsible company engaged in the manufacture of asbestos products would have been given pause by the kind of report that Dr. Grant received from Destefano in September of that year concerning the levels of asbestos dust in the Tyler factory. In fact, considering Dr. Grant's credentials, any other response would have been astonishing, for in addition to being medical director of the Pittsburgh Plate Glass Company and medical consultant to Pittsburgh Corning, he was a member of the American Medical Association, the American Industrial Hygiene Association, and the American Academy of Occupational Medicine, and would one day become president of the American College of Preventive Medicine.

In any case, in December of 1966 Dr. Grant paid a visit to Dr. George A. Hurst, clinical director of the East Texas Chest Hospital—a branch of the Texas State Department of Health that happens to be in Tyler—and asked him to conduct a medical survey of the workers at the Pittsburgh Corning plant to determine if they were encountering health problems as a result of their exposure to asbestos. Dr. Hurst immediately set about designing a study of the workers, which included physical examinations, questionnaires, X-rays, and pulmonary-function tests. On February 3, 1967, having received approval from his superiors at the Texas State Department of Health, in Austin, he wrote Dr. Grant that the study could be conducted at a cost to Pittsburgh Corning of forty-two hundred dollars, and that, upon Dr. Grant's approval, it would be started by the first of May and completed as soon as possible.

On March 7th, however, Dr. Grant wrote a letter informing Dr. Hurst that Pittsburgh Corning had decided to forgo the proposed study in favor of some studies that would be conducted by Dr. Lewis J. Cralley, who was associate program chief for field studies and epidemiology in the Public Health Service's Division of Occupational Health, in Cincinnati. Dr. Grant explained that the Public Health Service had been interested for some time in doing environmental and medical studies of the asbestos-products industry, and had recently agreed to include Pittsburgh Corning's plants in Tyler and Port Allegany in the environmental study. According to Dr. Grant, the Public Health Service did not then have sufficient funds to perform the medical study but hoped to receive additional money for that purpose in July. "For this reason, I would like to hold off until July on making a final decision on your proposed medical study," he wrote Dr. Hurst. "Our management is vitally interested in accomplishing the medical study but would like the USPHS to accomplish it as part of their total study. If USPHS can't do the medical study, they would like to consider your proposal further."

An environmental survey of the Tyler plant, which consisted of taking eighty-two samples of air in the factory, was conducted on March 20, 1967, by engineers sent there by Dr. Cralley. However, more than a year passed before Dr. Cralley's people got around to informing Pittsburgh Corning of the results of the survey. The report was dated March 27, 1968, and it was sent to J. W. McMillan, the works manager of the Tyler plant, with copies to Dr. Grant and Dr. Cralley. In many respects, McMillan must have found it a baffling document. On the one hand, it informed him that when twenty-seven of the air samples collected in his factory were analyzed by a standard method, dust concentrations exceeded the threshold limit value of five million particles per cubic foot "in a number of locations." On the other hand, it indicated that when the air samples were analyzed by a new method in use in Great Britain (and soon afterward adopted in the United States), asbestos-fiber counts were considered high in forty-four of the fifty-five other samples. (The fact was that in five of the samples the asbestos-fiber count was twenty to thirty times the range that was considered high by the British Occupational Hygiene Society. Moreover, with the exception of Pittsburgh Corning's plant in Port Allegany, where asbestos-dust levels were also very high, the over-all asbestos-fiber counts in the Tyler factory were far greater than those measured in any of some thirty other asbestos-products factories that had been surveyed by the Division of Occupational Health during the previous three years.) In spite of this, the report made no mention that a health hazard might exist at the Tyler plant, nor did it advise the works manager of the factory to improve the ventilation system or to institute better housekeeping practices—or, indeed, to correct any condition that might have led to the excessive fiber counts it described. Instead, the report concluded by telling him that "your cooperation in this study is sincerely appreciated and the data gained from your plant are of considerable value." It is not known whether the works manager had the benefit of any medical interpretation of the report. Nor is it known whether he had any other way of ascertaining the dimensions of the health hazard that existed in his factory. It is known, however, that two years later he died of mesothelioma.

Since the primary responsibility of the Division of Occupational Health was to protect workers from occupational disease, the omission from its report of any concern for the health of the workers at the Tyler plant seems puzzling, to say the least. Part of the trouble undoubtedly stemmed from the roundabout manner in which the Occupational Health people had to go about their business, for at the time they had no legal authority to enter and inspect factories and no enforcement power of their own. To gain access to factories, they had to be expressly invited by state departments of labor, or by the few state departments of health that had rights of access, or, as in the case of Pittsburgh Corning's Tyler plant and the other asbestos plants they were studying, by the companies that owned the factories. Since the Occupational Health people usually had to go hat in hand to industry in order to initiate their asbestos field studies, they appear to have felt a certain constraint about using the information they gathered. For example, in making arrangements to gain access to plants and take air samples, field-studies engineers invariably gave oral assurance to plant management that the identity of individual factories would be kept confidential and would be released only to the appropriate state agencies. In practice, however, the Division of Occupational Health almost never forwarded interpretations of the health consequences of its findings to state agencies, and in most cases it didn't even send them the sampling data—the report on the Tyler plant being no exception—and that, in effect, prevented any possibility of action to remedy any health hazards. The pledge of confidentiality, of course, precluded any possibility that the data collected in the surveys would be made known to the workers whose health was being affected or to the unions representing them. Moreover, in order not to embarrass management or make workers apprehensive, the government engineers who took air samples during the environmental surveys not only were forbidden to discuss the nature of their activities with any workers they encountered but were also instructed not to wear respirators, which would have afforded them some protection against the hazard of inhaling asbestos dust. As a further extension of this solicitous policy toward industry, the Occupational Health people were careful not to alarm management by reporting in writing the existence of health hazards in any of the asbestos factories they surveyed or by recommending improvements in ventilation equipment and housekeeping procedures to reduce the levels of asbestos dust. In short, they simply took air samples, analyzed them, and reported fiber counts, without drawing any inference as to what the fiber counts might mean in terms of the health and well-being of the men who were exposed to them.

In order to understand more fully what lay behind this practice, it might be helpful to examine the attitude toward asbestos disease held by the people who were in charge of the asbestos field studies at that time. As it happens, I spent several hours one afternoon in March of 1968—a week or so before the report on the Tyler plant was sent out—discussing the asbestos problem with Dr. Cralley and some of his associates in the Division of Occupational Health. At the time, I was looking into the biological effects of asbestos, and I had flown out to Cincinnati to see Dr. Cralley at the suggestion of Dr. Murray C. Brown, who was then chief of the Division of Occupational Health, with offices in Washington. Dr. Cralley, who received a

Ph.D. in industrial hygiene from the State University of Iowa, had joined the Public Health Service in 1941. At the time of our meeting, he had been in charge of epidemiology and field studies for the Division of Occupational Health for nearly four years. He was a fellow of the American Public Health Association, a past chairman of its Occupational Health Section, a past chairman of the American Conference of Governmental Industrial Hygienists, a member of the Committee on Asbestosis and Cancer of the International Union Against Cancer, and an adjunct assistant professor of environmental health at the University of Cincinnati.

At the beginning of our conversation, Dr. Cralley explained that the asbestos-field-studies program had two components—an environmental team, which had been taking dust counts in asbestos-textile and friction-materials plants since 1964, and a medical-epidemiological team, which would soon begin to give X-ray examinations, pulmonary-function tests, and blood tests to five thousand men who worked in these factories. He did not tell me that his engineers had already conducted environmental surveys of Pittsburgh Corning's insulation plants in Tyler and Port Allegany, perhaps because the initial focus of the field-studies program was into other areas of the asbestos-products industry, and these two surveys were exceptions. According to Dr. Cralley, the purpose of the field-studies program was to establish criteria for a possible lowering of the threshold limit value for asbestos. It would take many years to develop these criteria, however, for Dr. Cralley's medical studies of the asbestos-factory workers were designed to proceed from the time of first examination, rather than to reconstruct the events of the past, as Dr. Selikoff's and Dr. Hammond's study of the asbestos-insulation workers had done. Dr. Cralley explained that it would take from two to four years to complete the first medical examinations of the five thousand men he proposed to study, and that if funds were available the men would be re-examined every five years thereafter. "By following these men for the next fifteen or twenty years, we hope to establish a dose-response relationship for asbestosis," he told me. "Then we'll try to determine what level of exposure carries with it no discernible health hazard."

When I asked Dr. Cralley if this twenty-year-from-now evaluation would take into consideration the development of lung cancer and mesothelioma, he replied that it would not—that he was interested only in asbestosis. I then asked him about the medical studies indicating that mesothelioma could occur with minor exposure to asbestos, and he shrugged and replied that in his opinion the association between mesothelioma and asbestos was not proved.

At this point, Dr. William S. Lainhart, assistant chief of field studies in charge of the medical-environmental team, who was sitting in on our talk, explained that since the main purpose of the program was to trace the natural history of asbestos disease, little would be known about the incidence of lung cancer or mesothelioma until the five thousand men under study were re-examined in future years. "Ideally, we'd like to take a bunch of twenty-year-olds, put them into an asbestos plant where we know the exact dust levels, and observe them for the next fifty years, or until they die," he said. "Of course, we can't do that. We have to devise studies that are practical. For this reason, we estimate that it will take us from fifteen to twenty years to evaluate with any accuracy the medical effects of today's environment in the asbestos industry."

When I asked about the high rate of asbestosis, lung cancer, and mesothelioma that was already afflicting workers in the asbestos industry, Dr. Cralley told me that such diseases were the result of exposures sustained over the past twenty years or so, and that because great improvements had been made in ventilation systems and industrial-hygiene procedures in the meantime, he expected to find much less disease in the future. When I asked him what he would consider a high rate of disease in the men he proposed to examine, he replied that he would not care to estimate. "We'll have to come to that when we come to it," he said. "Remember that practically everyone is susceptible to chest disease to some extent, and that you can get chest disease even from digging in your garden. With the means we now have at hand, we can only *assume* that asbestosis and other diseases are related to asbestos exposure. Our first priority, therefore, is to study our five thousand men over a long period, and use our observations of what happens to them as the criteria for developing a new standard."

At the close of our meeting, I asked Dr. Cralley why, in view of the fact that asbestosis and cancer had afflicted great numbers of asbestos workers since the turn of the century, it had taken so long for the government to begin studying asbestos in earnest. Dr. Cralley said he didn't know. "All I know is that the first real interest came from industry," he told me. "They asked for our help back in 1964, and they have cooperated with us magnificently."

Since it was my understanding that the asbestos industry had never been particularly eager to have its operations scrutinized, I was surprised to hear this, and asked Dr. Cralley what segment of the industry had made the request for help and cooperated so magnificently. "The Asbestos Textile Institute," he replied.

Whatever interpretation one wishes to place upon the rather leisurely approach to the problem of asbestos disease by the administrators of the asbestos field-studies program, the fact remains that during the first seven years of its existence the program placed no emphasis at all upon control of dust levels in asbestos factories, or upon preventive measures for the workers who were exposed to the dust. Indeed, almost all the meaningful data about dust levels in asbestos factories which were acquired by the program between 1964 and 1971 simply accumulated in the files of its Cincinnati offices, as did all the data on the medical examinations of asbestos workers it conducted after 1968. For its part, the asbestos industry seems to have been quite content with this quiet state of affairs. On the one hand, it could state publicly that, with its assistance, the United States Public Health Service was investigating the possible hazards of industrial exposure to asbestos. On the other hand, it could rest assured privately that, because of the long-term nature of these studies, no information would be forthcoming for many years, and that, because of the pledge of confidentiality, none of it would find its way into the hands of anyone who might seek to remedy any hazards that were found in the meantime.

As for the Tyler plant, it seems to have been considered a kind of fluke by almost everyone concerned. By March 1968, the medical study of the workers in which Pittsburgh Corning was supposed to be vitally interested was either forgotten or held in abeyance. Dr. Grant never did anything further about the study that Dr. Hurst had designed, and Dr. Cralley and his associates in the Division of Occupational Health never got around to conducting their study. Meanwhile, as the health situation at the Tyler plant was going from bad to worse to appalling, a parade of government inspectors continued to troop through the place without any apparent awareness of the hazards that were staring them in the face. On February 13, 1969, still another safety-and-health inspection of the factory was conducted, this time by industrial-hygiene engineers from the Dallas regional office of the United States Department of Labor's Wage and Labor Standards Administration. The Department of Labor inspectors were authorized to enforce the industrial-health regulations of the Walsh-Healey Act of June 30, 1936, which had been amended to apply to companies holding federal contracts of ten thousand dollars or more. They found a number of unsatisfactory conditions in the plant, including, again, a substandard ventilation system and inconsistent use of respirators by the workers, and they proceeded to take air samples in six areas of the plant to determine the asbestos-dust levels. By that time, in spite of the foot-dragging of the Division of Occupational Health, the American Conference of Governmental Industrial Hygienists had proposed that the threshold limit value be lowered from five million particles per cubic foot (the standard that had been in effect for more than twenty years) to two million particles per cubic foot, the first of a series of downward revisions they were to consider—each an admission that previously recommended guidelines had allowed workers to inhale concentrations of dust now deemed harmful. (The two million particles were considered the equivalent of twelve asbestos fibers longer than five microns per cubic centimeter of air—five microns being one-five-thousandth of an inch, and a cubic centimeter of air being an amount equal to what might be contained in a small thimble.) The Department of Labor inspectors, however, not only had no equipment to measure asbestos-fiber counts in terms of the proposed conference standard but analyzed the air samples they took in terms of a standard applicable not to asbestos, a known carcinogen, but, rather, to nontoxic nuisance dusts, such as wood dust and chalk powder. As a result, they failed to realize that even the new standard for asbestos was being exceeded dozens of times over in the Tyler plant, and contented themselves with recommending that Pittsburgh Corning issue respirators to employees working in dusty areas of the plant. What seems especially ironic about this is that back in the forties, when UNARCO was operating its Paterson factory—the one upon which the Tyler plant was modeled—not only had it paid its employees five cents extra an hour to wear respirators, which they were obliged to do by insurance underwriters anyway, but it had also repeatedly pointed out to the workers that wearing respirators was a precaution that should always be taken in any asbestos factory, and had threatened to fire men who refused to wear them.

As for the inadequate ventilation system, the Department of Labor inspectors recommended in their report (which was sent to James H. Bierer, the president of Pittsburgh Corning, and to Charles E. Van Horne, who had recently become the manager of the Tyler plant) that the company "make a study of the present system with professional advisers and come up to standard, or present qualified proof that the present system is operating, within the minimum specified ventilating range." Instead of reinspecting the Tyler plant to make certain that the company had complied with these recommendations, however, the Department of Labor people simply took Pittsburgh Corning's word that approved respirators would be issued to its employees and that the ventilation system would be improved. Indeed, nobody from the Department of Labor visited the factory again until November of 1971.

For its part, Pittsburgh Corning asserts that by May of 1969 the wearing of approved respirators was required for all employees in the Tyler plant—an assertion that is denied by most of the men who worked there—and that the ventilation system had been duly studied with an eye to improving it. During 1969, the company did indeed engage the services of Dr. Morton Corn, professor of occupational health at the Graduate School of Public Health of the University of Pittsburgh, who visited the Tyler and Port Allegany plants and proposed some engineering controls to bring asbestos-dust levels in them down to recommended limits. But if any significant improvements were made in the ventilation system of the Tyler plant, they were clearly not sufficient to bring the dust levels within such limits. In January of 1970, engineers from Dr. Cralley's group at the Division of Occupational Health—which by then had become the Bureau of Occupational Safety and Health—returned to the factory and took seventeen air samples, and these showed the average airborne-asbestos level to be more than double the proposed twelve-fiber standard. In keeping with earlier practice, however, the engineers chose not to point out the existence of any possible health hazard in their report to Pittsburgh Corning, or to make any recommendations for lowering asbestos-dust levels in the Tyler plant. Nor did they forward their findings to the Texas State Department of Health or to any other agency with enforcement powers. Thus did the Bureau of Occupational Safety and Health add to the series of virtually meaningless surveys that had begun back in July of 1963, when Pittsburgh Corning engaged the Industrial Hygiene Foundation to evaluate the asbestos-dust hazard in the Tyler plant. During those six and a half years, five separate studies and inspections of the factory had been conducted, and more than a hundred samples of air had been gathered and transported to laboratories in various parts of the country, where they had been counted, weighed, assayed, and painstakingly analyzed by industrial hygienists who, depending on what standard they were using, had reported their findings in terms of dust particles per cubic foot or dust weight per cubic meter or fiber counts per cubic centimeter but never in terms of what the dust and fibers that the workers were inhaling might be doing to their health.

By this time, however, some preliminary data concerning the mortality experience of the men who had worked in the UNARCO factory in Paterson were being developed, and what the data revealed might have led one to anticipate a most unhappy fate for many of the workers at the Tyler plant. With the aid of a grant from the National Institute of Environmental Health Sciences, which had become concerned about the potential asbestos hazard to the general public, and wanted accurate data concerning it, Dr. Selikoff had set up an asbestos-control program in Paterson in 1968, and had begun to trace the sixteen hundred and sixty-four men who had been employed at the Paterson plant between 1941, when it opened, and 1954, when UNARCO closed it and transferred its operations to Tyler. This was a laborious process, for Dr. Selikoff and Dr. Hammond had only the names of the men to go on, and addresses for them that were from fourteen to twenty-seven years old. By January of 1970 they had managed to trace most of the nine hundred and thirty-three men who had worked at the factory between 1941 and 1945, and who, if they were still alive, had passed the twenty-year mark since their initial exposure to asbestos dust. They were also able to collect death certificates for almost all of the four hundred of these men who had died. As in the case of the insulation workers in New York Local 12 and Newark Local 32, the results were alarming, for, even though the study was incomplete, the death certificates showed an extraordinarily high frequency of death resulting from asbestosis, lung cancer, gastrointestinal cancer, and mesothelioma.

Meanwhile, Dr. Selikoff had continued to maintain a close watch on the insulation workers' health, paying particular attention to those men with more than twenty years' experience, whom he examined once or twice a year. He was thus able to detect symptoms of illness in many of these men much earlier than they might have been detected otherwise, and to observe with great accuracy the mortality experience of the three hundred and seventy men whom he had found in 1963 to be survivors of the original six hundred and thirty-two members of the union in 1942. The death rate continued to be disastrous. Between January of 1963 and March of 1968, sixty deaths would normally have been expected to occur among these men. Instead, there were a hundred and thirteen. Fifteen were caused by asbestosis, and, as with the earlier study, the rest of the excess turned out to have been caused by cancer of various kinds. Two or three bronchogenic cancers would have been normal, but twenty-eight occurred, and, where the actuarial tables predicted only two gastrointestinal cancers, there were eight. In addition, a sharp rise in the number of mesotheliomas was observed. In the earlier study, in which none had been expected, there had been four in two hundred and fifty-five deaths; between 1963 and 1968, thirteen of these rare tumors were found in a hundred and thirteen deaths.

In the spring of 1969, Dr. Selikoff presented these findings to the International Conference on Pneumoconiosis, in Johannesburg, South Africa, which was attended by dozens of scientists from all over the world—among them Dr. Brown, the chief of the Bureau of Occupational Safety and Health, who had been designated a vice-president of the conference, and Dr. Cralley, who had become director of the Bureau's Division of Epidemiology and Special Services. By this time, Dr. Selikoff and other scientists had developed reliable data to show that the insulation workers whose disastrous mortality experience he had documented had developed asbestosis or cancer by working in areas of asbestos dust whose levels were considerably below the twelve-fiber standard proposed by the American Conference of

Governmental Industrial Hygienists. Armed with this data, Dr. Selikoff recommended to the Bureau of Occupational Safety and Health in October of 1969 that an interim level for asbestos be set at two fibers per cubic centimeter—a standard that had also been recommended by the British Occupational Hygiene Society, and had been adopted by Her Majesty's Inspectorate of Factories. Dr. Selikoff pointed out that such a standard was intended only for the prevention of asbestosis, and not of cancer, which might well be associated with exposure levels below those resulting in asbestosis. By way of emphasizing this possibility, he pointed out that among the insulation workers there were as many excess deaths resulting from mesothelioma as from asbestosis, and twice as many excess deaths resulting from lung cancer. In spite of the alarming data supporting Dr. Selikoff's recommendation for a lower level, the bureau's asbestos experts appear to have been unconvinced that such action was necessary. A year later, when the bureau finally got around to proposing a standard for asbestos, it settled upon the twelve-fiber standard. As usual, the Bureau of Occupational Safety and Health was lagging far behind, for by this time—autumn of 1970—the Conference of Hygienists had proposed lowering the level for asbestos to five fibers.

To more fully comprehend the absurdity of such proposals and recommendations, one should know that there are always many fibers smaller than five microns in length in any amount of air containing asbestos dust. In fact, most experts in the field readily acknowledge that there may be hundreds, if not thousands, of these smaller fibers, or fibrils—tinier particles into which asbestos fibers readily fragment—simultaneously present for each one longer than five microns. (Indeed, if it were not for the electron microscope, the extent to which asbestos is fibrous would be difficult to believe, for approximately a million individual fibrils can lie side by side in a linear inch of chrysotile asbestos, whereas about four thousand glass fibers—such as those found in various insulation materials—or six hundred human hairs can be aligned along the same distance.) Little is known about the disease potential of fibers smaller than five microns in length; nor does anyone know how many asbestos fibers of any length must be inhaled in order to induce scarring of the lungs, cancer, and mesothelioma. Why, then, count only fibers longer than five microns—which, in effect, constitute only a tiny portion of the total? The reason is simply that the average industrial-hygiene laboratory is equipped with an ordinary phase-contrast optical microscope, capable of resolving only relatively large particles, whereas most particles smaller than five microns can often be seen only by electron microscopy, which is expensive and not readily available. Hence, even though recommended standards of two, five, or twelve fibers greater than five microns in length per cubic centimeter, or thimbleful, of air might actually reflect a hundred, or even a thousand, asbestos fibers per thimbleful, such standards have continually been justified on the basis of economic feasibility, sheer convenience, and wishful thinking—in other words, in the hope that counting only the larger particles would at least serve as an index for measuring the contamination of the air being studied. As for how these patently and admittedly inaccurate counts of fibers per thimbleful can be translated in terms of the lungs of asbestos workers, it should be pointed out that in a normal eight-hour working day a normal worker will breathe in and out about eight cubic meters of air. Since each cubic meter contains a million cubic centimeters, or a million thimblefuls, the worker is breathing in and out eight million thimblefuls of air each day. Thus, an asbestos worker toiling in an environment that is supposed to contain, say, only two fibers greater than five microns in length per thimbleful of air can in fact be inhaling anywhere from eight hundred million to eight billion asbestos fibers and fibrils of all sizes each day. No one knows for sure how many of these inhaled particles may subsequently be exhaled, but recent studies of the aerodynamics of asbestos fibers suggest that as many as fifty per cent of the fibers may well be retained in the lungs. Not that anyone needs aerodynamic studies to prove that the lungs will retain vast numbers of asbestos fibers. That has been proved beyond a doubt by Dr. Langer, of the Mount Sinai Environmental Sciences Laboratory, who, using electron microscopy to analyze lung-tissue specimens in autopsies of asbestos workers, has been able to calculate that as many as a hundred thousand billion to a million billion asbestos fibers and fibrils had accumulated over the years in the lungs of some of them. However, even as late as the autumn of 1970, no one in the Bureau of Occupational Safety and Health or, for that matter, in the independent medical community (let alone in the boardrooms of the asbestos industry) was talking about the hazard in terms of human lungfuls of literally billions upon billions of asbestos fibers. Everyone was talking about it, as almost everyone still is, in the euphemistic terms of thimblefuls of air containing two, five, or, at the very most, a dozen fibers. In this way did a few needles become the metaphor for—indeed, the medically and scientifically accepted definition of—a whole haystack.

At the end of 1970, however, an event occurred that showed some promise of overcoming the ignorance, laxity, and confusion that had so long enveloped the asbestos problem and other occupational-health problems. On December 29th, after two years of prodding from industrial unions, led by the United Steelworkers of America and the Industrial Union Department of the AFL-CIO, Congress passed Public Law 91-596—the first comprehensive occupational-health legislation it had enacted since the Walsh-Healey Act. Known as the Occupational Safety and Health Act of 1970, Public Law 91-596 sought to "assure safe and healthful working conditions for working men and women," and under its terms the federal government was authorized to develop and set mandatory occupational-safety-and-health standards applicable to any business that engaged in interstate commerce. The Secretary of Labor was given the authority to promulgate improved standards, and to enforce them by conducting inspections of factories and other workplaces and by issuing citations and imposing penalties if the standards were violated. The Department of Health, Education, and Welfare was made responsible for developing criteria for the establishment of the safety and health standards, including regulations for dealing with toxic materials and harmful physical agents and for instituting education and training programs to produce an adequate supply of manpower to carry out the provisions of the Act. So that the department could perform these functions, the Act provided for a National Institute for Occupational Safety and Health, called NIOSH, which replaced the Bureau of Occupational Safety and Health, and which was also given authority to enter factories for inspections and investigations, but, since the Act did not go into effect until April 28, 1971, and since NIOSH did not begin its operations until June 30th of that year, little or nothing was done during the next few months to resolve the problem of industrial exposure to asbestos. This delay was disheartening to many trade-union people and to independent medical researchers who had hoped for quick action on a new asbestos standard. However, the business-as-usual attitude that had characterized government policy toward the operations of the asbestos industry for so long was about to be shattered by a series of disclosures that, fittingly, would have their apotheosis in the revelation of the atrocious working conditions that had prevailed through the years at the Tyler plant.

The new turn of events got started on May 20, 1971, when industrial hygienists from the Meadville office of the Pennsylvania Department of Environmental Resources sent Pittsburgh Corning a report on some recent inspections they had made at the company's insulation plant at Port Allegany, a small town in the northwestern part of the state. The Commonwealth of Pennsylvania had not yet adopted the standard of twelve fibers per cubic centimeter, so the hygienists who inspected the Port Allegany plant were using the long-outmoded standard of five million particles per cubic foot, which is roughly the equivalent of thirty fibers per cubic centimeter. Even though Pittsburgh Corning had installed some new ventilating equipment in the factory during the previous two years, the inspectors found that dust levels exceeded the old standard in five of twenty-five air samples. In one of the samples, the count was more than twenty-six million particles, which meant that there may have been approximately a hundred and fifty fibers per cubic centimeter of air in that location—five times the outdated standard. Such dust levels—and even higher ones—had, of course, been found and ignored in the Tyler plant for years, and the management of Pittsburgh Corning may well have come to expect this process to be repeated elsewhere. The Pennsylvania inspectors, however, gave the company sixty days to improve the ventilation system and institute better housekeeping practices to reduce dust levels at the factory.

While Pittsburgh Corning's managers were mulling over this unexpected situation, some personnel changes were taking place at NIOSH that would soon cause them additional problems. Having reached retirement age, Dr. Cralley was about to leave the Division of Epidemiology and Special Services, which was being reorganized into the Division of Field Studies and Clinical Investigations. A number of positions in the new division had opened up, among them that of chief medical officer, and it was filled on July 1st, with the appointment of a young doctor named William M. Johnson. A native of Olean, New York, Dr. Johnson was brought up in Saranac Lake, graduated from the Stanford University School of Medicine in 1968, interned at the State University of New York at Buffalo, and had just completed a two-year training program in occupational health at the Harvard School of Public Health. He had decided to fulfill his military obligation by putting in a two-year stint with the United States Public Health Service, and, as things turned out, it did not take him long to become immersed in his job there. Within a few days of Dr. Johnson's arrival at the NIOSH offices in Cincinnati, one of the engineers who had been conducting field studies under Dr. Cralley told him about the environmental surveys of asbestos factories that had been in the files for several years. "That engineer was particularly concerned about the situation at the Tyler plant," Dr. Johnson has recalled. "And when I started digging through the files myself, I realized he had good reason to be, for it was plain as day that there was an incredibly serious health problem down there. At that point, I went to Dr. Cralley and asked him whom I should see about the situation. He suggested that I get in touch with Dr. Grant, Pittsburgh Corning's medical consultant. However, when I called Dr. Grant, on July 13th, he told me there really wasn't much of a health problem at the Tyler plant, because the place was so dusty that people didn't stay around there long enough to get sick. He also told me that there were no plans to improve the factory's ventilation system, and that the company planned to convert from asbestos to mineral wool in the near future."

Since Dr. Johnson had received considerable instruction at Harvard in the effects of asbestos exposure, he was less than reassured by his conversation with Dr. Grant. During the next two weeks, he gathered as much information as he could about conditions at Tyler and Port Allegany; then he discussed the situation thoroughly with Dr. Joseph K. Wagoner, who had arrived at NIOSH on August 1st to replace Dr. Cralley as director of the new Division of Field Studies and Clinical Investigations. Dr. Wagoner had received his Doctor of Science degree in epidemiology and biostatistics from the Harvard School of Public Health in 1970, and had previously spent ten years as an epidemiologist with the Public Health Service's National Cancer Institute, where he was instrumental in the long struggle to develop and institute standards for the protection of uranium miners, who were occupationally exposed to radioactive dust. He was as seriously disturbed about the potential health hazard at Tyler as Dr. Johnson was, and together the two men decided to make it and other asbestos-products factories their first order of business.

In the meantime, pressed by the deadline given by the Pennsylvania state inspectors for cleaning up the Port Allegany plant, and aware of Dr. Johnson's concern about conditions at the Tyler factory, Pittsburgh Corning made a move to head off some of the pressures that were building up against its asbestos operations. On August 3rd, the company filed an application for a variance from occupational-safety-and-health standards with the Assistant Secretary for Occupational Safety and Health of the Department of Labor, in Washington. Under the terms of the Occupational Safety and Health Act of 1970, the Secretary of Labor could grant a variance to an employer if he determined that the employer "has demonstrated by a preponderance of the evidence that the conditions, practices, means, methods, operations, or processes used or proposed to be used by an employer will provide employment and places of employment to his employees which are as safe and healthful as those which would prevail if he complied with the standard." Pittsburgh Corning's application for a variance was signed and submitted by E. W. Holman, the vice-president in charge of manufacturing and technology—who told the Tyler *Courier-Times* that he knew of no specific Pittsburgh Corning employee suffering from significant illness as a result of working with asbestos. In the application, Holman stated that the threshold limit value for asbestos was exceeded in some areas of the company's Tyler and Port Allegany plants, that the company had been unable to comply with the required standard because of the unavailability of effective ventilation equipment, and that the ineffectiveness of the available equipment had become particularly significant since the reduction of the standard for asbestos from twelve fibers per cubic centimeter to five. The fact that such a reduction in the official standard had not taken place but had only been published as a proposed change by the Conference of Hygienists suggests that Pittsburgh Corning's managers either were afflicted with a bad case of nerves or were trying to obscure the fact that neither factory was in compliance even with the obsolete standard of five million particles per cubic foot, let alone the twelve-fiber standard itself.

The application for a variance went on to say that as of June 1971, the company had spent nearly two hundred thousand dollars for research and development of a mineral-wool substitute for asbestos, that it would begin to use mineral wool in some of its operations in August, and that it hoped to make a complete conversion to mineral wool at the Port Allegany and Tyler plants by the middle of 1972. As for the steps the company had taken to provide working conditions as safe and healthful as those which would prevail if the government standard for asbestos had been complied with, Holman stated that Pittsburgh Corning had supplied approved respirators and had required workers to use them; that it had also purchased and was experimenting with new respirators; that it had provided dust-collection and ventilation apparatus; that it had expanded its program of periodic medical examinations; that it had improved housekeeping procedures by the more frequent use of vacuum cleaners; and that it had "provided and will continue to provide health education programs that fully explain to its employees the health hazards associated with asbestos exposure and how they can protect themselves." In a sworn affidavit attached to the application for a variance, Dr. Grant stated that he had knowledge of the matters set forth in the application "so far as said application states that the applicant has provided for its employees health education programs that explain to said employees the health hazards associated with asbestos exposure and how they can protect themselves."

Back at the Division of Field Studies and Clinical Investigations, in Cincinnati, several weeks were to pass before Dr. Johnson and Dr. Wagoner would learn of Pittsburgh Corning's application for a variance. Meanwhile, on August 9th, Dr. Johnson called Horace Adrian, chief of the industrial-hygiene program of the Texas State Department of Health, in Austin, and told him of the extraordinarily high dust levels that had been found in the Tyler plant. Adrian told Dr. Johnson that he had never seen copies of any inspection of the Tyler plant—indeed, he gave Dr. Johnson the impression that he did not know the factory existed—and he would look into the situation as quickly as possible. A week later, Adrian informed Dr. Johnson that he was going to Tyler the next day, August 17th, to conduct a walk-through survey of the factory and to meet Dr. Grant, who also planned to be there. (As it turned out, the purpose of Dr. Grant's visit to Tyler on August 17th was to give the workers there an educational talk on the health hazards of asbestos exposure, which Pittsburgh Corning had already claimed to have done in its application for a variance, filed two weeks before.) On August 24th, Dr. Johnson telephoned Dr. Grant, in Pittsburgh, to tell him that the Division of Field Studies and Clinical Investigations wanted to examine the workers at the Tyler plant. Dr. Grant replied that Dr. Hurst, of the East Texas Chest Hospital, had just finished giving the Tyler workers medical examinations, including X-rays and pulmonary-function tests, and he suggested that NIOSH might wish to defer its study of the men until the results of Dr. Hurst's tests could be made available.

In the light of this development, Dr. Johnson and Dr. Wagoner decided to hold off for the time being on their examination of the Tyler workers, and to conduct a medical survey of the men in the Port Allegany plant. (At that point, they had no idea that almost five years before Dr. Hurst had outlined a proposed—and rejected—medical study of the Tyler workers for Pittsburgh Corning.) On September 7th, in order to make arrangements for the survey of the Port Allegany workers, Dr. Johnson went to the plant, where he met Dr. Grant for the first time, and heard him give a talk to the workers on the health hazards associated with asbestos. According to Dr. Johnson, Dr. Grant indicated in his talk that the levels of asbestos dust at Port Allegany were not high enough to be considered dangerous to health. Dr. Grant also claimed that the dust levels were considerably lower than the ones that the insulation workers studied by Dr. Selikoff had been exposed to. (Actually, Dr. Selikoff had demonstrated that the insulation workers were exposed to levels of asbestos dust far below the twelve-fiber standard.) In addition, Dr. Grant implied that cigarette smoking was an important factor in the development of asbestosis, although such few data as are available indicate a very limited effect of cigarette smoking on lung scarring. (Dr. Grant may have misinterpreted some studies conducted by Dr. Selikoff and Dr. Hammond, which showed that asbestos workers who smoke cigarettes run eight times the risk of dying of lung cancer as cigarette smokers in general, and ninety-two times the risk of men who neither work with asbestos nor smoke.) But what Dr. Johnson found most disturbing of all was Dr. Grant's assertion that, in addition to smoking cigarettes, a man would have to undergo from twenty to thirty years of exposure to asbestos before experiencing any adverse effects. Later that day, Dr. Johnson made a point of telling Dr. Grant, in the presence of union officials, that radiological evidence of pulmonary fibrosis had been found in men with less than ten years' exposure, and that there was strong medical evidence to support the belief that lung cancer and mesothelioma could occur at exposure levels far below those that could cause asbestosis. This encounter with Dr. Grant seems to have marked a turning point in Dr. Johnson's dealings with Pittsburgh Corning, for, he has explained, "I came away from it feeling that Dr. Grant had grossly minimized the hazard of working with asbestos, and I assumed that he had probably done the same thing at Tyler on August 17th." Dr. Johnson's mistrust of the company's intentions was heightened a few days after this, when he discovered that Pittsburgh Corning had filed the application for a variance with the Occupational Safety and Health Administration; he believed the application to be not only self-serving but downright false in its claim that the company had undertaken to inform its employees adequately about the hazards of working with asbestos.

In addition, Dr. Johnson and Dr. Wagoner had other reasons to fear that no immediate action would be taken to reduce the health hazard at Tyler and Port Allegany. In the middle of August, they had prepared a memorandum expressing their concern about conditions at the two plants (it also included data from Dr. Cralley's files about excess mortality among workers in several large asbestos-textile factories) and sent it to Dr. Marcus M. Key, the director of NIOSH, which had set up its headquarters in Rockville, Maryland. During August and September, as it happened, there were huge internal problems at NIOSH headquarters about how the institute should carry out its role under the Occupational Safety and Health Act and how it should coordinate its activities with those of the Department of Labor's Occupational Safety and Health Administration, which had been given responsibility for enforcing health standards under the Act. An administrative crisis ensued, with the result that no one at NIOSH headquarters could give Dr. Johnson or Dr. Wagoner any assurance that something would be done to alleviate the conditions at the Tyler and Port Allegany plants. Feeling increasingly frustrated, Dr. Johnson and Dr. Wagoner decided after the encounter with Dr. Grant that the situation was serious enough to warrant their taking matters into their own hands. The following week, Dr. Johnson telephoned Steven Wodka, the legislative aide for the Oil, Chemical, and Atomic Workers International Union, and told him about the environmental studies of dust levels in the factory which he had found buried in the files. (The two men had met previously to discuss the problem of workers exposed to beryllium at the Kawecki Berylco Industries plant, in Hazleton, Pennsylvania, where, as in the case of Tyler, the Bureau of Occupational Safety and Health had gathered data about health hazards associated with a substance that the workers were exposed to but had for a number of years neglected to inform the workers of the dangers involved.) Upon learning of the situation at Tyler, Wodka immediately sent Dr. Johnson a letter requesting that he make the environmental data available, and, on September 24th, Dr. Johnson sent them off to the union's Legislative Department, in Washington.

When Wodka discussed the Tyler situation with his boss, Anthony Mazzocchi, the director of the Legislative Department, Mazzocchi remembered Dr. Selikoff's telling him about a study that he, Dr. Hammond, and Dr. Churg were conducting of the mortality experience of the men who had been employed at the Paterson plant. Mazzocchi quickly got in touch with Dr.

Selikoff, who, as it turned out, had completed the first part of the study a week or two before, and was about to present his data at the Fourth International Pneumoconiosis Conference of the International Labour Organisation, in Bucharest, on September 29th. When Mazzocchi told him about the environmental data on the Tyler plant that Dr. Johnson was making available to the union, Dr. Selikoff sent Mazzocchi the results of his study of the Paterson workers. They were as alarming as the mortality data on the asbestos insulators. Of three hundred and thirty-three men who had been employed at the Paterson factory for a year or more between 1941 and 1945, eighty-eight had died by December 31, 1959, and fifteen could not be traced. However, Dr. Selikoff, Dr. Hammond, and their associates had managed to trace every one of the remaining two hundred and thirty men who were alive on January 1, 1960, and had studied their experience up to June 30, 1971. Using the standard mortality tables, Dr. Hammond calculated that no more than forty-seven deaths would normally have been expected to occur among these men during that eleven-and-one-half-year period. Instead, there were a hundred and five. Fourteen of the deaths were caused by asbestosis, and, as with the insulation workers, a large majority of the excess deaths were caused by cancer. Two or three lung cancers would have been normal, but twenty-five occurred, and deaths from cancer of the stomach, the colon, and the rectum were three times what the standard mortality tables predicted. In addition, although none would normally have been expected, there were five deaths from mesothelioma.

Mazzocchi and Wodka were profoundly disturbed at the results of the Paterson study, for they could only conclude that the similarity of operations in the two factories meant that much the same thing would happen to the men at Tyler. Meanwhile, as they were trying to decide what to do, Pittsburgh Corning was informed by the Department of Labor that no action would be taken on its application for a variance from occupational-safety-and-health standards until after a public hearing, and that no hearing could be held until the spring of 1972. Since this meant that the company would be forced to comply with existing health standards until then, and since the company had failed to prove that mineral wool could be successfully substituted for asbestos in high-temperature pipe covering, the Pittsburgh Corning people found themselves in a bind. In the early part of October, therefore, they told representatives of Local 4-202, in Tyler, with whom they were negotiating a new contract, that they might have to shut the plant, but said that they wished to consider the local union's proposals on wages, health, and safety before making a final decision. When Mazzocchi and Wodka learned of the company's action, they suspected that Pittsburgh Corning was using the threat of a shutdown to force the local union to minimize its demands for improved working conditions at the Tyler plant. After consulting with representatives of Local 4-202, the two men determined that a strict-compliance program should become part of the union's contract proposal on safety and health. They then submitted a formal request to Dr. Johnson at NIOSH, on October 7, 1971, for a comprehensive industrial-hygiene study of the Tyler factory and a medical survey of the men who were working there. Upon receiving this request, Dr. Johnson made the necessary arrangements with Pittsburgh Corning and with the Texas State Department of Health, which had previously offered to cooperate, and a survey of the plant was scheduled to take place in the last week of the month.

As things turned out, the NIOSH inspection of the Tyler plant coincided with a crescendo of protest that had been building up for many months over the plight of tens of thousands of workers throughout the country who were being exposed to excessive concentrations of asbestos dust. On August 3rd—the day Pittsburgh Corning filed its application for a variance—Dr. Selikoff wrote to James D. Hodgson, the Secretary of Labor, making carbon copies for leading officials of six labor unions whose members worked with asbestos, and for Dr. Key at NIOSH. His letter said:

Dear Mr. Hodgson:

Your department has published initial standards in the Federal Register, in accordance with the Occupational Safety and Health Act of 1970. I understand that these are "initial" and that modifications may be expected as a result of research criteria being developed by the National Institute for Occupational Safety and Health.

One "standard" in the published list is so wrong, and represents such serious hazard to workmen, that I advise its urgent revision.

I refer to the standard for asbestos, which would allow workmen to be exposed to environments containing as many as twelve fibers per cubic centimeter of air. Our research in one asbestos trade—insulation work—demonstrates that work in the past in areas with levels of two to three fibers per cubic centimeter of air has resulted in a very great increase of death due to cancer and to asbestosis. Just how serious this has been may be appreciated from current statistics: at present, one in every five deaths among insulation workers is due to lung cancer, one in ten to cancer of the pleura or peritoneum, one in ten to scarred lungs or asbestosis.

The proposed level is much higher than actually *now* exists. It is so high as to make totally ineffective current efforts by both industry and labor to control this unhappy occupational health hazard.

In Great Britain, the approved level is *less than* one-fifth the standard here proposed and levels of twelve fibers per cubic centimeter for more than even ten minutes would be sufficient to require that the workman wear protective clothing and use an efficient respirator.

Mr. Albert E. Hutchinson, President of the International Association of Heat and Frost Insulators and Asbestos Workers, AFL-CIO, has calculated that there are approximately one hundred thousand men employed doing asbestos insulation work in the United States, in various unions and in various industries. Utilizing statistical calculations by Dr. E. Cuyler Hammond, director of the Department of Statistics of the American Cancer Society, it may be predicted that, if the situation remains the same and does not improve, there will be more than seventeen thousand excess deaths of lung cancer among these men, as well as almost ten thousand unnecessary deaths of cancer of the pleura or peritoneum, ten thousand wholly preventable deaths of asbestosis, and many thousand other cancer deaths, in this one trade alone. Thousands of deaths will occur in other industries, to add to the unhappy toll of this serious error.

I urge you, then, to recall this standard, and substitute one that will help protect workingmen forced to work with this dangerous material.

Although Dr. Selikoff's letter did not engender any immediate response from Secretary Hodgson, it did evoke profound concern among the union officials to whom he sent carbon copies. When Dr. Selikoff returned to New York from Bucharest, Sheldon W. Samuels, the director of Health, Safety, and Environmental Affairs for the AFL-CIO's Industrial Union Department, invited him to attend a meeting of the IUD's *ad hoc* Committee on the Asbestos Hazard, in Washington, on October 18th, so that he might present additional information, which the union people hoped would enable them to get some effective action from the Department of Labor on the problem of occupational exposure to asbestos. Meanwhile, having discovered additional reports in the old Bureau of Occupational Safety and Health files which showed excessive dust counts in a dozen more asbestos factories (there was also an incomplete study of mortality among employees of those factories, which showed an extraordinary number of deaths resulting from asbestosis among men in their forties and fifties), Dr. Johnson and Dr. Wagoner continued to express concern about the asbestos problem to their superiors at NIOSH, who they hoped would take a firm stand in advising the Secretary of Labor to promulgate a tough emergency standard for asbestos. On October 4th, feeling that the situation was getting out of hand, Dr. Johnson and Dr. Wagoner visited Dr. Selikoff in New York to exchange information about the asbestos problem in general and, in particular, about Johns-Manville's asbestos-textile factory in Manville, New Jersey, where that corporation has owned and operated the largest complex of asbestos-products factories in the world for more than fifty years. Dr. Cralley's files had yielded up several environmental studies showing that excessive dust levels had existed in the Manville asbestos-textile factory at least since 1965, when the first study was made; a 1969 medical survey showing findings consistent with asbestosis in thirty-one of a hundred and seventy-nine chest X-rays of the factory employees; and an incomplete mortality study showing, on preliminary analysis, four deaths from mesothelioma and at least ten other asbestos-related deaths among a hundred and eighty asbestos-textile workers. Indeed, the situation at Manville appeared to be similar to the one at Tyler, and on a larger scale. In 1967, engineers from Dr. Cralley's office had taken air samples that were not analyzed for asbestos fibers until September of 1971, when Dr. Johnson discovered the data in the files. That September, too, Dr. Johnson had fiber counts completed on over a hundred air samples that engineers from Dr. Cralley's division had taken at the Manville asbestos-textile plant during the spring of 1971. The fiber counts showed that even then there were as many as twenty fibers per cubic centimeter of air in some operations of the plant.

Dr. Johnson and Dr. Wagoner believed these data to demonstrate a serious and persistent health hazard at the Johns-Manville factory, and they were anxious to know if Dr. Selikoff could give them any additional information, particularly with regard to other cases of mesothelioma that might have occurred in Manville. As it happened, Local 800 of the United Papermakers and Paperworkers Union, which represented the company's production workers, had provided Dr. Selikoff with a roster of its membership several months before, and he, Dr. Hammond, and one of their associates at Mount Sinai, Dr. William J. Nicholson, had just begun a mortality study of the Johns-Manville employees, so Dr. Selikoff was able to give his visitors details on about a dozen deaths among the workers which had been caused by mesothelioma.

On October 5th, Dr. Johnson and Dr. Wagoner went to

Trenton, where they met with the commissioner of the New Jersey Department of Labor and Industry and the deputy commissioner of the state's Department of Health, and told them of the health hazard that they believed to exist at the Johns-Manville plant. When they asked these officials to investigate the situation, however, they learned that the state considered it to be a federal problem and, in any case, did not possess modern fiber-counting equipment for such a task. The following day, the two men took the data they had compiled on the Johns-Manville textile factory to the New York Regional Office of the Occupational Safety and Health Administration, in New York City, only to discover that the people there did not possess adequate fiber-counting equipment, or even know how to use such equipment properly. With that, they flew back to Cincinnati and got in touch with officials of the United Papermakers and Paperworkers Union. A few weeks later, they learned something from the union people that the commissioner of the New Jersey Department of Labor and Industry had known for more than a year—that in 1969 alone the Johns-Manville Corporation had paid out $887,341 in workmen's compensation to two hundred and eighty-five employees of the Manville plant who had become disabled with asbestosis.

At his meeting with union leaders in Washington on October 18th, Dr. Selikoff documented his contention that the Department of Labor should establish a standard for occupational exposure to asbestos to replace the current twelve-fiber standard. At the same time, he stressed the fact that any standard for asbestos exposure could be concerned only with the prevention of pulmonary fibrosis, and that little was known as to how low a standard might have to be in order to prevent asbestos-related cancer, which had accounted for fully three-quarters of the excess deaths among the insulation workers.

On November 4, 1971, the IUD transmitted through George Taylor, the executive secretary of the AFL-CIO's standing committee on safety and occupational health, a letter urgently requesting Secretary Hodgson to use the powers granted him by the Occupational Safety and Health Act of 1970 to declare an emergency standard governing the industrial use of asbestos. The letter declared that the existing twelve-fiber standard constituted "a license to jeopardize without effective restraint the lives of millions of workers," and urged the Secretary to declare an emergency standard of two fibers per cubic centimeter and to issue a bulletin prescribing that an appropriate label be affixed to each container of asbestos and asbestos products warning workers of danger. In addition, the letter asked the Secretary to get in touch with the administrator of the Environmental Protection Agency "to enable him to investigate the necessity for invoking the imminent-danger provisions of the Clean Air Act, as amended in 1970, to protect our families and communities from the effects of ambient asbestos that escapes from the workplace."

As might be supposed, the IUD's letter placed considerable pressure upon Secretary Hodgson to take some kind of action. When he did so, however, on December 7th, he declared an emergency standard for asbestos of five fibers longer than five microns per cubic centimeter of air. This, of course, was an emergency standard only in the eyes of the Department of Labor, since it was two and a half times as great as the standard requested by the IUD, and since the Conference of Hygienists had already published it as a proposed change. It is not known what, if any, medical data prompted the Secretary of Labor to select the five-fiber standard, or why he chose to disregard the data indicating that asbestos disease could occur at this level of exposure. Perhaps he was seeking a middle ground that he hoped would be satisfactory both to industry and to the union people. If so, he was neglecting the responsibility placed upon him by the Occupational Safety and Health Act for promulgating standards that, even if they entailed conflict, would assure "the greatest protection of the safety or health of the affected employees." In any case, it is a pity that he was not aware of the NIOSH report on Pittsburgh Corning's Tyler plant, which was then slowly making its way through the bureaucratic labyrinth, for the story of the plant constituted incontrovertible evidence of the sorry tangle of ignorance, laxity, and lack of communication that had from the very beginning characterized government policy toward occupational exposure to asbestos. The story of Tyler also made a mockery of one of the basic assumptions behind this policy: that the government could and would force industry to abide by a numerical fiber standard, and, by so doing, could insure healthful working conditions in asbestos factories.

The NIOSH inspection, which was conducted between October 26th and October 29th, included an industrial-hygiene survey, carried out by engineers from NIOSH's Division of Technical Services, and a medical survey, performed by a three-man team from the Division of Field Studies and Clinical Investigations. The medical team was headed by Dr. Johnson, who has a vivid memory of his first look at the Tyler plant. "Two carloads of us drove in from Dallas late on the afternoon of the twenty-sixth," he recalls. "The factory was situated in an industrial district on the outskirts of town, and it consisted of a pair of wood-shell buildings, each of which was about a thousand feet long, fifty feet wide, and thirty feet high. When we arrived, we were met in the front office by Mr. Charles E. Van Horne, the plant manager, and since it was late in the day, there was just time for a quick preliminary walk-through. The place was an unholy mess. Why, compared with it, the Port Allegany plant looked like a hospital operating room! A thick layer of dust coated everything—from floors, ceilings, and rafters to drinking fountains. As we walked through the interior, we saw men forking asbestos fiber into a feeding machine as if it were hay. They obviously had no idea of the hazard involved. Farther down the line, we came upon some fellows with respirators hanging around their necks, who were sitting in an open doorway eating watermelon. I hate to think of the fiber counts on those slices of watermelon. I remember turning to Dr.

Richard M. Spiegel, one of my assistants. 'This is intolerable,' I told him. He was as shocked as I was."

It did not take Dr. Johnson and his associates from NIOSH long to realize that at virtually every stage of the manufacturing process enormous quantities of asbestos dust were being spewed out into the factory. In addition to poor housekeeping procedures, the chief cause was a grossly inadequate ventilation system. Other aspects of the plant's operation were found to be equally hazardous. The scrap-grinding machine, where refuse from various operations was made reusable, was extremely dusty and lacked sufficient ventilation equipment, and a fan near the feed hoppers simply contributed turbulence that redispersed dust into the working environment. Moreover, both the scrap grinder and the feeding machines relied on a dust-collection system that consisted of canvas bags inside the plant, beneath the roof. These bags were periodically emptied by mechanical shaking, and when this happened huge amounts of asbestos dust were released into the air; then, after it had settled, instead of being vacuumed the dust was swept into piles with push brooms. The ventilation equipment on the saws in the finishing department was also found to be inadequate; excessive amounts of asbestos dust were escaping into the air there as well.

Because of Pittsburgh Corning's application for a variance, the NIOSH inspectors learned, the wearing of respirators had been mandatory in all areas of the plant since August. However, instead of being used for emergency or backup protection, as industrial-hygiene standards prescribed, the respirators were obviously being employed in the Tyler plant as substitutes for an adequate dust-control system and for proper housekeeping. Nor did the company have any adequate program for selecting, fitting, cleaning, and maintaining the respirators worn by its employees, and many of the men were wearing them improperly. In addition to noting these hazards, the inspectors saw that the factory's lunchroom was within fifty feet of one of the dustiest operations in the plant, and that workers were allowed to enter it wearing clothes that were contaminated with asbestos. The NIOSH men also discovered that compressed-air outlets throughout the plant were being used to blow excess dust off the employees—a practice that simply reintroduced asbestos fibers into the working environment.

As a result of these multiple deficiencies in the ventilation system and in the operating procedures of the Tyler plant, Dr. Johnson and his associates were not surprised to find that, of a hundred and thirty-eight air samples taken at different locations in the factory, a hundred and seventeen exceeded the recommended five-fiber standard. (Of course, even if the Tyler plant *had* observed that standard, workers not wearing respirators would still have been inhaling at least forty million asbestos fibers in an eight-hour working day.) What astonished them, however, was how *much* the levels exceeded the recommended standard, at almost every step in the manufacturing process. In the mixing department, where the feed hoppers and the scrap grinder were situated, the maximum concentration was a hundred and eighty-nine fibers per cubic centimeter of air, and the average was seventy-five. In the forming department, where the material was rolled on mandrels, the maximum concentration was a hundred and thirty-four fibers per cubic centimeter, and the average was thirty-nine. In the finishing department, where pipe covering was trimmed and sawed, the maximum concentration was two hundred and eight fibers per cubic centimeter—an approximation, for the air sample was actually too dusty to permit an exact count under a microscope—and the average was forty-one. Even in the inspection and packing department, where the finished product was weighed, boxed, and shipped, the maximum concentration was ninety-two fibers per cubic centimeter, and the average was twenty-three—nearly double the interim standard of the Department of Labor, nearly five times the standard recommended by the hygienists, and ten times the level of exposure that Dr. Selikoff and other epidemiologists had found to be responsible for the extraordinary number of excess deaths among the insulation workers. But the true intensity of the exposure of the Tyler workers can only be appreciated when one recognizes that such concentrations of long asbestos fibers per thimbleful of air really meant that, before they were required to wear respirators in August, some of these men were inhaling up to a billion of the longer fibers each working day, and many more of the shorter ones, which were not being counted by the engineers.

Since asbestos-induced cancers generally take at least twenty years to develop, and the Tyler plant had been in operation for only seventeen years, Dr. Johnson and his associates did not yet expect to find neoplasms among the sixty-three men working there. However, in order to complement the X-rays and pulmonary-function tests that had been performed in August by Dr. Hurst, of the East Texas Chest Hospital, they examined the employees for rales—crackling sounds in the chest which can occur with asbestosis—and for finger clubbing, a thickening of tissue at the fingertips which often occurs with asbestosis. After comparing notes with Dr. Hurst, they determined—even without the benefit of the X-rays, which they were not allowed to see—that seven of the eighteen workers with more than ten years of employment at the factory met at least three of four criteria for asbestosis. (These criteria included, besides rales and finger clubbing, dyspnea, which is shortness of breath, and marked reduction of forced vital capacity, which is an inability to take sufficient air into the lungs because of pulmonary fibrosis.) Reduced pulmonary function was also observed in some workers who had been employed at the plant for less than five years. Because of these findings, Dr. Johnson and his associates concluded that the health of the sixty-three employees at the Tyler plant had been gravely jeopardized, but they wished to have the X-rays reviewed by an expert panel of radiologists before making a definitive diagnosis of asbestosis on an individual basis. As things turned out, however, they were seeing only the tip of the iceberg, for when they got around to examining the company's employment records they discovered that a total of eight hundred and ninety-five men had worked in the plant at one time or another. Considering the disastrous mortality figures of the men who had worked at the Paterson factory between 1941 and 1945, this was disturbing news, to say the least. It also provided a chilling corollary to Dr. Johnson's first conversation with Dr. Grant, back in July, when Dr. Grant had suggested that there wasn't much of a health problem at the Tyler plant, because people didn't work there long enough to get sick.

The preliminary report of the NIOSH survey, declaring that an extremely serious occupational-health situation existed at the Tyler plant, was sent, on November 16th, to Dr. James E. Peavy, the commissioner of the Texas State Department of Health, and copies went to a number of other officials, including Dr. Key, Dr. Grant, Van Horne, and John K. Barto, the regional administrator for the Occupational Safety and Health Administration, in Dallas. Barto received the report on November 18th, and acted quickly on it, for he was aware that industrial hygienists from the Department of Labor's Dallas office had inspected the Tyler plant nearly three years before, and, even though they had found an inadequate ventilation system and faulty respiratory protection, had not taken any effective action to remedy the situation, or even reinspected the factory to see whether Pittsburgh Corning had corrected the defects. On November 23rd, Barto sent Clarence R. Holder, his assistant, and John P. Boyle, an industrial hygienist, to conduct still another inspection of the Tyler plant. As might be expected, their findings simply substantiated those of the survey conducted by NIOSH. On December 1st, Holder and Boyle informed Van Horne that violations of occupational safety and health regulations found in the plant included not only improper wearing of respirators but failure to examine workers to determine whether they had the physical capacity for wearing respirators, inadequate housekeeping, and insufficient dust control. They also told the plant manager that citations would be issued and penalties imposed, and that the violations would require immediate corrective action—except extensive improvements in ventilation and dust equipment, for which a later date would be set. Holder and Boyle then returned to Dallas, where in the next two weeks, as they awaited analysis of air samples they had taken, they wrote up a lengthy report of their inspection, which was sent to Holman, the corporation's vice-president in charge of manufacturing and technology, at Pittsburgh Corning's home office, on December 16th. In the meantime, Secretary Hodgson had declared the emergency five-fiber standard for asbestos, but since the inspectors for the Occupational Safety and Health Administration had surveyed the factory under the twelve-fiber standard published in the *Federal Register* of May 29th, they decided it would be unjust to apply the new regulation *ex post facto* to the situation at Tyler. As things turned out, their sense of fair play made little difference, for the obsolete twelve-fiber standard was exceeded—in some instances, ten times over—in forty-two of the forty-four air samples they had taken at the plant. Even so, the Occupational Safety and Health people failed to cite Pittsburgh Corning for violating any fiber standard, or even to list, in the citation for insufficient dust control, the specific dust levels they had found. Moreover, although their *Compliance Operations Manual* clearly compelled them to follow Section 17 (k) of the Occupational Safety and Health Act, which states that "a serious violation shall be deemed to exist in a place of employment if there is a substantial probability that death or serious physical harm could result from a condition which exists . . . unless the employer did not, and could not with the exercise of reasonable diligence, know of the presence of the violation," the inspectors listed the conditions they had discovered at the Tyler plant not under the heading of "Serious Violations" but under the heading of "Nonserious (Other) Violations," which, according to their manual, applied to situations "where an incident or occupational illness resulting from violation of a standard would probably not cause death or serious physical harm." If the violations had been considered serious, the Administration could have assessed Pittsburgh Corning as much as a thousand dollars for each one. For nonserious violations, the Occupational Safety and Health Administration could have assessed a penalty of anywhere from a thousand dollars each to no penalty at all, depending upon the inspector's judgment of "the severity of the injury or disease most likely to result." Given this latitude, the Administration people undertook to grant Pittsburgh Corning the benefit of every doubt. For three violations—improper wearing of respirators, failure to examine the workers to see if they could wear respirators, and inadequate housekeeping—they proposed a fine of twenty-five dollars each. For insufficient dust control, they proposed a fine of a hundred and thirty-five dollars. The total came to two hundred and ten dollars.

There has since been considerable speculation in government circles, among trade unions, and within the independent medical and scientific community about the lenient attitude adopted by the Occupational Safety and Health Administration toward Pittsburgh Corning after the November survey. Apologists for the Administration claim that because the Secretary of Labor announced the emergency standard for asbestos after the survey was conducted harsher penalties might have been interpreted as unfair harassment of the company. Some people say that, with the Administration in the process of redefining safety and health standards for workers, the regional inspectors in Dallas had received no clear guidelines from Washington about over-all policy and, fearful of exceeding their authority, were simply trying to muddle through an administrative vacuum. Other observers feel that the Administration was reluctant to invoke the letter of the law in the case of the Occupational Safety and Health Act, because to have done so would have opened a Pandora's box of health violations in other industries, which the Administration was ill-equipped to handle, since it had only forty industrial hygienists to enforce standards in some four million workplaces. And still others, including Mazzocchi, of the Oil, Chemical, and Atomic Workers International Union (OCAW), believe that the Administration's people were highly embarrassed by the NIOSH report on the Tyler plant, which pointed up the Department of Labor's failure to enforce the standards of the Walsh-Healey Act at the factory after a department inspection on February 13, 1969. Indeed, Mazzocchi believed that the Administration's people were hoping that if they slapped Pittsburgh Corning lightly on the wrist with some nonserious violations the whole affair would blow over.

Whether the truth lies in any one of these theories or, as seems more likely, in a combination of them, it became apparent in December of 1971 that the Tyler affair was not going to blow over. That it did not, as so many other occupational-health scandals had, was largely the result of the efforts of Mazzocchi, who was determined to make it a *cause célèbre.* For several years, he and his associates in the union's Legislative Department, aided by Dr. Glenn Paulson, a young scientist working under Professor René Dubos, at the Rockefeller University, had been holding conferences for factory workers around the country to point out the hazards existing in the industrial environment and to give the workers guidance in how to deal with them. It was a considerable undertaking, for the hundred and eighty thousand members of the union were being exposed to thousands of potentially toxic substances. Indeed, in certain factories men were working with more than five thousand different chemicals, yet federal standards existed for fewer than four hundred and fifty of them. Since Mazzocchi found that the lack of government standards was accompanied by lax government enforcement of existing health regulations and an attitude of almost total callousness on the part of industry, he became convinced that industrial workers were victims of a conspiracy designed to suppress knowledge of occupational-health hazards and to delay the passage and enforcement of laws to cope with them. A blunt-spoken man in his middle forties, with a fiery temperament, piercing eyes, and an unruly mane of black hair, he had begun to brood over what he considered the gross immorality that attended the plight of men who were either dying or being disabled early in life as a result of exposure to substances such as asbestos, whose adverse health effects had long been known to and ignored by a medical-industrial complex consisting of company doctors, industry consultants, and key occupational-health officials at various levels of the state and federal governments. Consequently, when the situation at Tyler was brought to his attention by Dr. Johnson in September of 1971, Mazzocchi saw it as, in his words, "the epitome of what was wrong with the way our country was dealing with the problem of industrial disease." However, he decided against making any public disclosure of conditions at the Tyler plant until the NIOSH survey had been conducted and written up. "I wanted to marshal all the facts and then choose the forum that would enable me to use the example of the Tyler plant to best effect, and to bring help to the unfortunate men who were working there," he has said. "I wanted the whole country to know in detail what had happened at that factory, and to understand that what had gone on there—the fruitless Bureau of Occupational Safety and Health inspections, the lack of enforcement by the Department of Labor, the company's disregard for the men, the whole long, lousy history of neglect, deceit, and stupidity—was happening in dozens of other ways, in hundreds of other factories, to thousands of other men across the land. I wanted people to know that thousands upon thousands of their fellow citizens were being assaulted daily, and that the police—in this case, the federal government—had done nothing to remedy the situation. In short, I wanted them to know that murder was being committed in the workplace, and that no one was bothering about it."

The forum Mazzocchi chose was the annual meeting of the American Association for the Advancement of Science, which was held at the Sheraton Hotel in Philadelphia on December 26th, and was attended by some fifty thousand scientists from all over the country. Along with Ralph Nader and Dr. Sidney Wolfe, the director of Nader's Health Research Group, Mazzocchi addressed eleven hundred of the delegates at a symposium entitled "Workers and the Environment," which was moderated by Professor George Wald, the Nobel Prize-winning biologist from Harvard. Nader and Dr. Wolfe spoke about occupational-health problems in general, and Mazzocchi zeroed in on Tyler as the quintessential example of the ignorance, neglect, and subterfuge that characterized government policy toward industrial hygiene. The complete and final report of the survey made by NIOSH had been sent out on December 21st, and he was thus able to document his charges with devastating effect. Mazzocchi was hoping that his disclosures about conditions at the Tyler plant would be picked up by newsmen covering the meeting and relayed to the nation. As things worked out, however, press coverage of the symposium tended to emphasize the more general aspects of occupational health that had been discussed by Nader.

In the meantime, notwithstanding the mildness of the penalties that accompanied them, the Administration citations had caused considerable consternation to the management of Pittsburgh Corning by setting March 31, 1972, as the date for completing extensive improvements in the ventilation- and dust-control systems of the Tyler plant. Combined with the failure of a company plan to use mineral wool as a substitute for asbestos in the manufacture of high-temperature pipe insulation, and the union's insistence that there be strict compliance with the recommendations of the NIOSH report, the deadline meant that the company would at long last be forced to clean up the plant or close it. Faced with this choice, Pittsburgh Corning's board of directors met in the middle of December and decided to shut the factory within a few months—after using up as much raw-asbestos stock as possible—in order to avoid further penalties.

The decision to close the Tyler plant rather than to spend the necessary time and money to make it safe obviously appealed to the managers of the company as the best way out of their difficulties. However, it did not solve the problem faced by Dr. Grant. Dr. Grant appears to have felt that his professional reputation had been called into question by the findings of the

NIOSH survey. On December 13th—five months to the day after he had told Dr. Johnson that the health situation at Tyler wasn't serious—Dr. Grant visited Dr. Johnson at the Division of Field Studies and Clinical Investigations to discuss the medical condition of the seven Tyler workers who showed signs and symptoms of asbestosis and, as Dr. Grant put it, "to avoid further misunderstanding" between NIOSH and Pittsburgh Corning. According to Dr. Johnson, the meeting was somewhat strained, for Dr. Grant was attempting to minimize the asbestos problem at the Tyler factory. "He tried to imply that our medical findings on the seven men were based more on epidemiological data than on clinical examination," Dr. Johnson recalls. "I replied that rales, finger clubbing, dyspnea, and reduced pulmonary function were well-recognized clinical evidence of asbestosis. Furthermore, I told him that back in October Dr. Richard Spiegel, of our division, was given to understand that some of the X-rays taken by Dr. Hurst showed evidence of pulmonary fibrosis. I also reminded Dr. Grant that in 1970 a manager of the plant had died of mesothelioma. Dr. Grant then informed me that the company was willing to release Dr. Hurst's X-rays to us—they had been made at Dr. Grant's request—but only on the condition that we come back to Tyler early in January, so that we could sit down and review them together. I got the distinct impression that he was trying to make it appear that NIOSH and Pittsburgh Corning had been working jointly on the problem from the very beginning. However, we wanted to get our hands on those X-rays, so I agreed to the meeting, even though I saw little reason for it."

On January 5, 1972, Dr. Spiegel and Richard A. Lemen, an epidemiologist who had also been a member of the original NIOSH survey team, returned to Tyler, and the following morning they met with Dr. Grant, Dr. Hurst, and William Farkos, Pittsburgh Corning's director of personnel and industrial relations, in a conference room at the East Texas Chest Hospital. The five men reviewed the medical reports on all of the employees—results of X-rays and pulmonary-function tests done by Dr. Hurst, and results of examinations for rales, finger clubbing, and dyspnea done by the NIOSH team—and concluded that seven of the employees had asbestosis. Then Dr. Hurst announced that his superiors at the Texas State Health Department would not release the chest films to NIOSH. But Dr. Grant said that the release would be acceptable to Pittsburgh Corning, and it was decided that he and Dr. Spiegel would fly to Austin to try to get permission from the Texas State Health Commissioner.

By arrangement with Pittsburgh Corning, Steven Wodka, of the OCAW, arrived in Tyler on the afternoon of that day to tour the plant and to meet with members of the local union committee. Wodka was accompanied on his tour by Herman Yandle, the chairman of the local committee; by Farkos; and by Van Horne, the plant manager; he found conditions in the factory to be much as they had been described in the NIOSH report. Huge piles of loose asbestos fiber lay on the floor of the plant, and visible clouds of dust were erupting from several operations. After the tour, Wodka—who had learned only the previous day that men from the Occupational Safety and Health Administration had inspected the plant in November—asked to see the citations they had issued, pointing out that the company was required to post them publicly for the information of the workers. When Van Horne and Farkos produced them, they professed to be ignorant of this requirement, although each citation sheet states that "a copy of the enclosed citation(s) shall be prominently posted in a conspicuous place at or near each place a violation referred to in the citation occurred." They also assured Wodka that the items marked for immediate compliance had been taken care of. Then they told him that Pittsburgh Corning planned no extensive changes in the ventilation- and dust-control systems by the March 31st deadline, and that the factory would be shut down sometime in the middle of February. At the same time, they rejected a request by Wodka that in the meantime the company provide in-plant facilities for storing respirators and for changing work clothes, so that the employees would not continue to contaminate their cars and homes with asbestos dust and thus expose members of their families to the risk of incurring asbestos diseases.

That evening, Wodka held a meeting with several members of the local union committee to discuss the health situation at the plant. At the outset, he informed them of the results of the mortality study Dr. Selikoff and Dr. Hammond had made of men who had worked at the Paterson factory, which obviously had frightening implications for the men who were employed at Tyler. "It was a somber moment, for it was the first time that they had been told how serious their situation was, and how precarious it would remain in the future," Wodka recalls. "However, they accepted the news stoically. They even seemed relieved to know the facts. You see, they had been living in doubt ever since the NIOSH survey, and still hadn't been informed that any of them had been diagnosed as having asbestosis."

The next morning, a final meeting was held at the East Texas Chest Hospital. In addition to Dr. Grant and Dr. Hurst, those present were Farkos and Van Horne; Dr. Spiegel and Mr. Lemen; and Wodka and Yandle. Dr. Grant began by saying that when he first visited the Tyler plant, in 1965, he had realized that a potential health problem existed there, and had done everything in his power to have the situation corrected. He took credit for instituting a number of surveys and studies of the plant's ventilation equipment, and blamed the Bureau of Occupational Safety and Health and Dr. Cralley, who had been in charge of field studies and epidemiology for the bureau, for not following through on a proposed medical study of the workers. Dr. Grant also insisted at the meeting that asbestos-dust levels at the Tyler plant were always within the legally specified limits, and he claimed that respiratory protection was made mandatory for all workers in the factory as early as May of 1969—a contention that was disputed on the spot by Yandle, who had worked in the plant for more than ten years, and who declared that, with a few exceptions, the wearing of respirators was not required until the summer of 1971. For his part, Dr. Spiegel listened to Dr. Grant's assessment of the history of the plant in silence. "I was not there to argue with him about what had or had not happened, or why," he recalls. "I was there to reiterate NIOSH's medical conclusions about the workers we had examined during our survey, and to bring back the X-rays taken by Dr. Hurst in August. In my opinion, however, Dr. Grant was trying to absolve himself and the Pittsburgh Corning Corporation of all responsibility for the conditions that had existed in the Tyler plant over the years, and for the lack of attention paid to the health of the workers. His speech sounded like a legal brief more than anything else. As I sat listening, I kept wondering why on earth he had not kept calling Dr. Cralley to insist that the bureau undertake its proposed medical studies, or, failing that, why he didn't get back to Dr. Hurst sooner than he did. Surely, in all the years since 1965 he could have found *someone* to examine the workers."

After Dr. Grant's opening summary, he and Dr. Spiegel formally announced their joint conclusion that at least seven of the eighteen workers with more than ten years of employment in the Tyler plant had symptoms and signs consistent with asbestosis. Dr. Grant and Farkos then said that Pittsburgh Corning would pay for the treatment of any disease that was occupationally related, provided that the treatment was performed at the East Texas Chest Hospital, but they were vague about whether the company would pay for any long-term medical followup of the sixty-three workers then employed at the plant. There was also no discussion whatever of any medical followup of the eight hundred and thirty-two men who had previously been employed at the Tyler plant, and who would carry the asbestos they had inhaled in their lungs for the rest of their days, presumably because no one could be held legally responsible for what might happen to them.

After the meeting, Dr. Spiegel and Dr. Grant flew to Austin to obtain permission for the release of the X-rays. The chest films had been an item of dispute between NIOSH and Pittsburgh Corning from the beginning; now the Texas State Department of Health people were apparently concerned about being in the middle of such a dispute, and they may well have begun to entertain second thoughts about the propriety of Dr. Hurst's having accepted a contract from Pittsburgh Corning to examine workers for occupational illness at one of the department's hospitals, when the department itself had never seen fit to conduct a health inspection of the Tyler plant during the plant's entire existence. By that time, practically everyone who had been involved in the Tyler affair seemed to be trying to cover his tracks.

While Dr. Spiegel and Dr. Grant were en route to Austin, Wodka and Lemen stayed behind to conduct a thorough investigation of Pittsburgh Corning's practices of dumping asbestos waste in the vicinity of the plant and of selling burlap bags contaminated with asbestos dust to local nurserymen. (Both practices had been mentioned in the NIOSH report, which informed Pittsburgh Corning and the Texas State Department of Health that they presented a serious health hazard, and recommended that all asbestos waste and used burlap bags be buried.) That afternoon, Wodka, Lemen, and Yandle drove out to a group of dumps in an inhabited area near a sewage-treatment plant, about a mile north of the factory. One of the dumps was the size of half a football field, and was covered to the depth of at least two feet with asbestos scrap. Thick vegetation sprouting through a covering of asbestos in surrounding gullies showed that the dump had been in use for many years. Yandle told Wodka and Lemen that asbestos scrap was still being trucked there daily, and his statement was substantiated by the presence of wads of asbestos fiber covering the access roadway. Since the company had undertaken to put soil over two dumps adjacent to the factory which had been seen by the NIOSH investigators during their October survey, its continued use of this third, more remote dump site could be construed only as an act of blatant disregard and deception. However, for sheer brazenness the Pittsburgh Corning people outdid themselves a few weeks later, when they learned that the third dump had been discovered. At that point, they simply posted an armed guard at the site, with instructions to turn all unauthorized persons away.

After taking photographs of the dump, Yandle, Wodka, and Lemen drove to the Eikner Nurseries, a huge outfit on the west side of Tyler, where they interviewed Elbert Fenton, the man in charge of sales and distribution. Fenton told them that a shipment of thirty thousand burlap bags, costing twelve hundred dollars, had been received from Pittsburgh Corning some months before, and he took his visitors out to a barn where some fifteen hundred of the bags were stacked in piles. The bags were visibly contaminated with asbestos, and even as Lemen and Wodka were telling Fenton about the hazard of using them, a woman worker entered the barn, picked up a bag, and, after shaking it vigorously to remove the dust, slung it over one shoulder and walked out. With Fenton's permission, Wodka and Lemen took two sample bags away with them in sealed containers for laboratory analysis; then, bidding Yandle good-bye, they drove back to Dallas and flew home. Meanwhile, Dr. Grant and Dr. Spiegel had met with the State Health Commissioner, who gave permission for the X-rays of the Tyler workers to be mailed to NIOSH headquarters in Cincinnati.

At the beginning of 1972, the Tyler affair was like one of those simmering quarrels between small nations that threaten to disrupt the balance of power in a sensitive geopolitical area where a number of larger nations have interests. The Oil, Chemical, and Atomic Workers saw it as an outright test of the Occupational Safety and Health Administration's willingness to enforce the provisions of the Occupational Safety and Health Act; the NIOSH people, who had exposed the situation at the factory to begin with, were waiting to see how the Occupational Safety and Health people would dispose of the recommendations that had been made to correct it, in order to gauge the effectiveness of the advisory role they themselves had been given under the Act; and the Occupational Safety and Health people, who had not seen fit to tell the union about the minor nature of the penalties they had imposed upon Pittsburgh Corning, could not have failed to view the affair as a potentially embarrassing scandal, coming at a most inopportune moment. Under the provisions of the Occupational Safety and Health Act, the emergency standard for asbestos that had been declared by the Secretary of Labor on December 7th had to be replaced by a permanent standard within six months. The law also required the Department of Labor to hold public hearings before it promulgated a new standard, and notice that hearings on the asbestos standard would begin in March was to be published in the *Federal Register* on January 12th. Since the permanent standard on asbestos was the first ruling the Occupational Safety and Health Administration would make under its mandate to redefine occupational-safety and health regulations, industry and labor were prepared to look upon the ruling as an indication of how determined or easygoing the Administration would be when it came time to set new standards for other hazards, such as lead, beryllium, carbon monoxide, mercury, silica, ultraviolet light, and cotton fibers. The public hearings on asbestos, then, loomed as a crucial test of strength between the independent medical and scientific community and the medical-industrial complex. On the one hand, the NIOSH people were preparing to send the Administration a document recommending that the permanent standard for occupational exposure to asbestos be set at two fibers per cubic centimeter, which was the standard that the British Inspectorate of Factories—the Administration's English counterpart—had set in 1968, the one that Dr. Selikoff had advocated in 1969, and the one that the AFL-CIO's Industrial Union Department's *ad hoc* Committee on the Asbestos Hazard had urged upon the Secretary of Labor in November of 1971. On the other hand, the major asbestos companies, led by the giant Johns-Manville Corporation, were preparing testimony to show that the five-fiber standard would prevent disease from occurring among asbestos workers, and economic statistics to demonstrate that a two-fiber standard would drive them out of business. In order to weigh all the evidence and decide upon a safe level of exposure to asbestos, the Occupational Safety and Health Administration would obviously have to conduct itself in a completely impartial manner, for, even under the best of circumstances, its decision was bound to be controversial. As things stood, however, its claim to impartiality was in question, because of its failure to enforce even the standards for asbestos that it was now compelled to consider obsolete. And since its failure had nowhere been more recently and nakedly apparent than in its handling of the critical situation that existed in the Tyler plant, the Administration was unhappy about the possibility that the citations and penalties it had sent Pittsburgh Corning in December might become public knowledge. The fact that under the provisions of the Occupational Safety and Health Act such information was public property seems to have been forgotten.

Already stung by the disclosures Mazzocchi had made at the science meeting, and eager to avoid future embarrassment, the Occupational Safety and Health Administration now sought to place itself above reproach by publicly revising its attitude toward occupational exposure to asbestos. On January 4th, George C. Guenther, who was the Assistant Secretary of Labor and the director of the Occupational Safety and Health Administration, held a press conference in Washington to announce the start of a target health-hazards program designed to control five toxic substances that were in widespread use in American industry—silica, lead, carbon monoxide, cotton dust, and asbestos. At this time, a special health-hazard fact sheet on asbestos was distributed to newsmen (and, presumably, to the Administration's regional administrators and inspectors), informing them that more than two hundred thousand workers across the nation risked serious disease from direct occupational exposure to asbestos and that asbestos fibers were being found in the lungs of people who had had no industrial exposure to the mineral at all. The fact sheet reminded everyone that the emergency standard for asbestos was an eight-hour average of five fibers greater than five microns in length per cubic centimeter of air, and it stated unequivocally that any exposure exceeding this level would henceforth be considered a serious violation of the Occupational Safety and Health Act.

It soon became apparent, however, that it was one thing for the director of the Occupational Safety and Health Administration to announce a new policy on asbestos in Washington and quite another for that policy to be put into effect by his subordinates in the field. On January 13th, a week after Farkos and Van Horne had assured Wodka that all conditions on which the Administration had called for immediate compliance had been corrected, the Tyler plant was reinspected by Clarence Holder and John Boyle, of the Administration's Dallas office. Holder and Boyle found on their second inspection of the plant that, contrary to recommendations made after their first visit, the workers had not been examined to determine whether they were physically able to wear respirators, that required improvements on a block saw and on feeding-machine enclosures had not been made, and that cleanup procedures in the feeding and building areas of the plant remained inadequate. The two men reviewed the situation with Van Horne, but when they stressed the necessity of cleaning up asbestos waste around the feeding machines and the builder units, the plant manager asked them whether this meant that the factory would have to be kept as clean as a dairy.

In spite of the company's reluctance to take this inspection seriously, Holder and Boyle issued no new citations but simply sent Pittsburgh Corning a notification of failure to correct three of the original violations and a list of proposed additional penalties. Moreover, in spite of the new policy on asbestos that had been announced by Assistant Secretary Guenther ten days earlier, they still chose to consider the violations at the Tyler plant nonserious. Therefore, the proposed additional penalties were based upon a schedule of a hundred dollars for each day of noncompliance, plus fifty per cent of the original fine. The period of noncompliance ran from December 21, 1971, to January 13, 1972, and the total additional assessment amounted to six thousand nine hundred and ninety dollars.

The notification of failure to correct violations and the list of new penalties were mailed to E. W. Holman, Pittsburgh Corning's vice-president in charge of manufacturing and technology, on January 17th. It is not known whether Holman received them prior to his telephone interview with the Tyler *Courier-Times* on January 19th, nor is it known whether he had read the citations that were mailed to him on December 16th; whether he had been informed of two NIOSH reports—one describing the critical occupational-health situation at the Tyler plant and the other the symptoms of asbestosis in seven of the workers there—that were sent to Dr. Grant, in November and December; whether he had been told by Dr. Grant of the death of Van Horne's predecessor, J. W. McMillan, from mesothelioma; or whether he was aware of the admissions about the incidence of asbestosis among the Tyler workers which Dr. Grant had made in Dr. Hurst's office, on January 7th. If Holman did not know of these things, he must surely have been one of the most ill-informed vice-presidents in corporate history. Whatever the case, on January 19th he told the *Courier-Times* that he knew of no Pittsburgh Corning employee suffering any significant illness as a result of working with amosite-asbestos insulation.

By that time, with the Tyler affair not just bubbling but about to boil over, practically everyone involved in the situation was maneuvering for position. The Texas State Department of Health had already put some distance between itself and the stove; Pittsburgh Corning was backing toward the door; the Occupational Safety and Health Administration had little choice but to sit on the lid; and the union and NIOSH were separately gathering fuel for the fire. On the very day that Holman talked to the *Courier-Times*, Lemen sent Wodka a report containing the results of tests performed by Roy M. Fleming, an engineer in the Environmental Investigations branch of NIOSH's Division of Field Studies and Clinical Investigations, on one of the burlap bags that had been taken from the nursery in Tyler on January 7th. The bag, still in its sealed container, was placed in front of a chair in a room that was free of asbestos. A man wearing a respirator and two dust samplers entered the room and sat on the chair. Taking the burlap bag out of the plastic container, he turned it inside out; then, standing, he shook the bag for sixty seconds. Next he sat down, spread the bag on his lap, and wrapped it around the base of a small tree. After tying the corners of the bag with wire, he placed the tree on the floor, stood up to brush off his clothes, and left the room. During the four minutes this operation required, the average concentration of asbestos fibers longer than five microns around the worker was four hundred and ninety fibers per cubic centimeter of air—ninety-eight times the emergency five-fiber standard and more than forty times the old twelve-fiber one. (Indeed, had he not been wearing a respirator, the man would have inhaled some four million fibers during the four-minute period.) In a covering letter that was sent with a copy of the report to the Texas State Department of Health on January 17th, Lemen and Fleming described the use of burlap bags that had contained asbestos as a potentially serious health problem, and, by way of documenting this charge, pointed out that a recent report in the *British Medical Bulletin* referred to cases of mesothelioma in women who had been engaged in cleaning such bags.

In the meantime, Wodka had been trying without success to get information from John K. Barto, the regional administrator of the Occupational Safety and Health Administration, in Dallas, on the Administration's actions about Tyler. Finally, on January 28th, Barto told Wodka that he was under orders from his superiors in Washington "not to disclose anything on the Tyler case to anybody." He did, however, tell Wodka about the followup inspection made by Holder and Boyle on January 13th, and about the additional proposed penalties levied against Pittsburgh Corning as a result of it. Barto refused to release the report of the Department of Labor's 1969 inspection of the Tyler factory or of the dust counts that Holder and Boyle had taken there in November, claiming that he felt obligated to withhold them because of the possibility that Pittsburgh Corning might contest the fines and send the Tyler case into litigation. Since Wodka and Mazzocchi considered this a lame excuse designed to cover up the Department of Labor's failure to implement the provisions of the law, they now determined to make a fresh attempt, as they put it, to "blow the lid off the Tyler affair and focus national attention upon it."

The two union officials were well aware that their decision amounted to a declaration of open war upon the Occupational Safety and Health Administration and Pittsburgh Corning. Indeed, for several weeks Wodka had been compiling a history of the Tyler plant for a confidential report entitled "Occupational Health Tragedy," which amounted to a battle plan for the approaching hostilities. In the concluding section of the report, Wodka urged that a press conference be held in Washington to provoke the Environmental Protection Agency into taking action on the open-air dumps and contaminated bags, to force the Occupational Safety and Health Administration to declare an imminent-danger situation at the Tyler factory under the provisions of the Occupational Safety and Health Act, and to denounce Pittsburgh Corning for its disregard of the health of the Tyler workers. Realizing that much of the opprobrium for what had gone on in the Tyler plant was bound to be offset by Pittsburgh Corning's decision to close the plant, Wodka concluded his report by saying:

> We have a responsibility to point out what protection Pittsburgh Corning failed to provide for the workers at Tyler and, instead, what utter contempt this company displayed for the lives of those workers, their families, and that community. The Tyler case is unique in that this company was aware as early as 1963 that something was wrong in Tyler. Pittsburgh Corning was not ignorant, but was fully aware of what was going to happen to those men in that plant. The company would very much like to see the plant shut down, the men quietly fade away, and the Tyler pin removed from their corporate map. The last place the company wants to be is on page 1.

The press conference that Wodka had requested was held on February 10th at the National Press Building in Washington, and was conducted by Mazzocchi, who opened the proceedings by reading telegrams that had been sent on February 8th by A. F. Grospiron, the president of the OCAW, to William Ruckelshaus, the Administrator of the Environmental Protection Agency, and to Assistant Secretary Guenther. The telegram to Ruckelshaus urged him to take immediate action regarding the open asbestos dumps in Tyler and the tens of thousands of contaminated bags that had been sold to nurseries in the Tyler area, and to place Pittsburgh Corning under surveillance to prevent the repetition of such practices in other places. The telegram to Guenther pointed out the Department of Labor's laxity in not having enforced asbestos regulations at the Tyler plant in the past, and urged him to investigate Pittsburgh Corning's operations in other places.

Mazzocchi next reviewed the chronology and the findings of the various inspections of the Tyler factory over the years, and then he launched into an angry denunciation of the callousness and ignorance that had allowed the conditions there to go uncorrected. "Industry doctors and government officials who cover up vital life-and-death information about worker health are guilty of the My Lai syndrome," he declared, and he went on to note that what had happened at Tyler was probably happening in hundreds of other plants in America, and was symptomatic of the occupational-disease problem in this country. He concluded by describing Pittsburgh Corning's reaction to the NIOSH survey and the Administration citations. "This company was receiving a subsidy in terms of years of men's lives," he said. "When that subsidy was terminated by belated government intervention, the company decided to quit the asbestos business altogether." Neither Guenther nor Ruckelshaus responded to the telegrams sent by Grospiron, and only two major newspapers ran articles on Mazzocchi's charges.

Down in Tyler, however, the Pittsburgh Corning plant got back into the news by way of a dispatch from United Press

International that was based upon a press release issued by union headquarters, in Denver, on February 9th. The release was concerned chiefly with the contaminated burlap bags. As a result, when the story appeared in the Tyler papers, the *Courier-Times* and the *Morning Telegraph*, on February 17th, there was no mention whatsoever of the NIOSH report, of the Administration citations and fines, or of the fact that serious illness had already manifested itself among the sixty-three-man work force at the Tyler factory. The report in the *Morning Telegraph* was carried beneath a banner headline that read "Threat of Cancer Alleged," the article in the *Courier-Times* appeared beneath the caption "Company Denies Claims by Union," and both pieces consisted mostly of charges leveled by a union official about the contaminated burlap bags and denials of the charges by Gerald J. Voros, a public-relations spokesman for Pittsburgh Corning. According to both accounts, however, Voros admitted that "workers at the plant had been given physical examinations to determine if any of them exhibited contamination symptoms," and that "those whom the examinations revealed had been affected were referred to their personal physicians." He was quoted as saying, when asked by the *Telegraph* what the company would do if any of these afflictions could be attributed to asbestos dust, "I don't know. We'll just have to wait and see." Voros was, of course, being noncommittal only about Pittsburgh Corning's responsibility to the sixty-three workers who were employed at the Tyler plant when it was shut down, on February 3rd, since neither Pittsburgh Corning nor UNARCO can be held legally accountable for what may happen to the eight hundred and thirty-two other men who toiled in the factory at one time or another during the seventeen years it was in operation.

Thus, in the middle of February, Pittsburgh Corning found itself compelled for the first time to issue public denials of wrongdoing in connection with its Tyler plant, even as it took steps to deny public access to the factory and its adjacent dumps, and to bury as much of the equipment and machinery from the place as possible. That the company should be placed on the defensive over some dumps and burlap bags, and not because of the awful jeopardy in which it had put the health of hundreds of its workers and their families, seems ironic. It should not, however, come as any surprise. Much of industry in the United States has long operated on the assumption that it could endanger the lives of its employees with relative impunity—and without embarrassing publicity and possibly damaging repercussions—so long as it did not overtly threaten the health and safety of the community at large. Underlying this assumption is the further assumption that workers are not so much a part of the community as part of the equipment and machinery of production. As such, upon being proved defective they become expendable. They can be replaced or transferred, or, if worst comes to worst, given workmen's compensation (which in most states is minimal) and retired. At that point, they cease to be anyone's responsibility. Like the eight hundred and ninety-five men who worked in the Tyler plant over the years, they are out of sight and out of mind. In a sense, therefore, like much of the factory itself, they are buried.

PART TWO

That Dust Has Ate Us Up

I first heard about the Tyler plant when I attended the press conference that Anthony Mazzocchi held in Washington, D.C., on February 10, 1972. Later, at my request, he provided me with copies of all the inspection reports, surveys, and studies that had been made of the factory over the years, and after studying them I decided to go to Texas and talk with some of the people who had been involved in the whole sad affair. On Monday, March 6th, I flew to Dallas, where I paid a call on John Barto, the regional administrator for the Occupational Safety and Health Administration. Barto is a heavyset, bejowled man in his early fifties. I was ushered into his office shortly after noon, and when we had shaken hands I took a seat at a conference table in front of the desk and explained why I had come to see him.

"Look," he said. "NIOSH referred the Tyler situation to us, and we confirmed what NIOSH had discovered. It's as simple as that."

I told him that I was familiar with the contents of the NIOSH report and with the results of the Occupational Safety and Health Administration's subsequent inspections, and that I was particularly interested in knowing more about the inspection of the factory that the Dallas office of the Department of Labor had conducted on February 13, 1969.

Barto replied that he had not arrived at the Dallas office until July of 1970, so we would have to discuss the matter with Clarence Holder, the assistant regional administrator, who had been there since 1961, and he buzzed Holder on an intercom and asked him to bring a copy of the 1969 inspection report. While we waited, Barto told me that when he took over as regional administrator there had been no industrial hygienist in the office and there had been only three inspectors for the region, a five-state area comprising Texas, Louisiana, Oklahoma, Arkansas, and New Mexico. "A plant as small as Tyler simply couldn't be checked under those conditions," he said. "Believe me, that was a minuscule problem compared to others we had."

Holder came in and took a seat opposite me at the conference table. He struck me as being an archetypal Texan—a tall, lean, tanned man in his late forties, with wavy blond hair and a rugged, seamy face. After Barto had explained the purpose of my visit, I asked Holder if the Department of Labor had ever inspected the Tyler plant before February 13, 1969. He shook his head. I then asked him if he knew of any previous inspections that had been conducted by agencies with enforcement power, such as the Texas State Department of Health. He replied that he did not. When I told him that I was particularly interested in learning more about the 1969 inspection conducted by the Dallas office, he replied tersely that on that date the Tyler plant had been found to be unsatisfactory according to the provisions of the old Walsh-Healey Act. I then asked Holder on what specific grounds the factory had proved to be unsatisfactory, and he answered that the exhaust systems were deficient in terms of airflow, and that there was excessive dust.

At that point, I inquired what steps had been taken to determine whether the Pittsburgh Corning people had corrected the unsatisfactory conditions in the Tyler plant. Holder replied that the company had agreed that it would initiate a research-and-development program to improve the ventilation system, that it would study the possibility of replacing asbestos with a substitute material, and that it would provide respirators in the meantime, to protect the workers in the particularly dusty areas of the plant. He went on to say that the Walsh-Healey Act allowed the company to exceed the asbestos-dust standard if respirators were issued and worn, and that the issuance of respirators had therefore satisfied him as far as any violations of the Act were concerned.

I asked Holder how he knew that the employees in the Tyler plant had received respirators.

"Because the company had to issue them in order to come into compliance with the Walsh-Healey Act," he replied.

Rephrasing the question, I asked him if he had any first-hand knowledge that the company had complied.

For a moment, Holder remained silent. Then, quietly, he asked me if, by any chance, I was hard of hearing.

I replied that I didn't think so but that perhaps I had not understood him.

"Well, I think you must be hard of hearing, fella," he continued, "because didn't I just explain all that to you?"

I repeated that perhaps I had not understood him, and, rephrasing the question once more, asked him how he could be certain that Pittsburgh Corning had issued respirators to its employees following the 1969 inspection. After glancing at Barto, Holder said that the company had assured him at that time that it would undertake a study to improve the ventilation system and that it would issue respirators. He had no reason to believe that Pittsburgh Corning had not kept its word, he added.

I asked Holder if it was normal procedure to accept a company's statement of intention as evidence of compliance with occupational-health regulations, and he replied that he felt it was.

I then asked him if he or anyone else from the Dallas office had ever reinspected the Tyler plant to determine whether the company had come into compliance.

At that, Barto held up his hand. "No sense beating around the bush with this," he said. "What Holder's saying is that we did not reinspect. Let's leave it at that."

Since this seemed the only sensible course to take, I got to my feet, thanked both men for their time, and left the room. Holder followed me out, and engaged me in conversation as I was putting on my coat. "You know, the trouble with respirators out there is some of those boys really like their chewing tobacco," he told me in a confidential tone. "Why, you can't get them to wear a mask no way."

I asked Holder if he knew that asbestos inhalation could cause not only pulmonary scarring but lung cancer, mesothelioma, and other malignancies. He replied that he had never heard of mesothelioma and that he doubted if there was any real proof that asbestos could cause cancer.

"If such proof existed, would you still characterize the violations at Tyler as nonserious?" I asked.

"You know, this health business is always being exaggerated, in my view," Holder said. "I didn't see any serious danger in the Tyler situation, and I still don't. How can there be a serious danger if it doesn't hurt you right away? I mean, how can it be called serious if you can go right on working with it?"

Half an hour later, I was driving a rented car east on Interstate 20, toward Tyler, about a hundred miles away. For the first fifty or sixty miles, the country was flat and dotted with grazing cattle; then it became hilly and wooded. Here and there, a pale-green wash of buds on bushes growing in wet draws signaled the approach of spring, but the hint seemed lost against a larger background of brown fields, skeletal trees, red clay soil, and a vast sky filled with circling buzzards. At about four-thirty in the afternoon, I turned off Interstate 20 and drove south on U.S. Highway 69 to a motel on the outskirts of Tyler, where I had made a reservation. At the turnoff, there was a large billboard, erected by the Peoples National Bank of Tyler, that read "Life Is a Bed of Roses."

When I got to the motel, I telephoned Dr. George Hurst, superintendent of the East Texas Chest Hospital, and made an appointment to see him the next afternoon. Then I got in touch with Herman Yandle, the local union chairman, who lives in Hawkins, a small town eighteen miles north of Tyler. Yandle has no telephone, but his wife's grandmother's house is next door, and she went over to get him. I had called him from New York and told him I wanted to talk with him, and when he came to the phone a few minutes later he said he would be along within an hour. While I waited for him, I drew up a list of questions for Charles Van Horne, Pittsburgh Corning's plant manager, whom I planned to see the following day.

Yandle, who arrived shortly before six o'clock, proved to be a tall, heavyset man with an engaging grin and an easygoing manner. He came into the room, stuck out a hand, and plopped down in an easy chair. I asked him to tell me something about himself and his work history, and he said that he was thirty-six years old and had gone to work at the Tyler plant in the autumn of 1961. He had spent three and a half years in the production department, he went on, working on the feeder and scrap machines, and for the past seven years he had been in the shipping department. The first time he had ever worn a respirator was in July of 1971, he told me, and it was also the first time he had ever seen any of his co-workers wearing respirators except for a short period in the spring of 1969, when two or three men working in very dusty areas had requested them. "It was more or less voluntary in 1969," he explained. "And after a few weeks the guys stopped wearing them."

I asked Yandle when he had first been told that asbestos was hazardous to work with, and he said it was in August of 1971, when Dr. Lee Grant came down to give the men a talk. "We were real mad about the masks," he said. "Heck, I'd been working there ten years without one, and they were a pain in the neck to wear, especially in the summer, when it got so hot in there you'd sweat and couldn't get enough air through to breathe. So when Dr. Grant called my shift into the shipping supervisor's office, the first thing we asked him was how long we were going to have to wear the respirators. He said no longer than a year, because the company was going to shift over from asbestos to mineral wool. He had a blackboard set up in there with a lot of complicated words and numbers on it that none of us could make head or tail of. I remember he said that the company could make the whole plant dust-free but they couldn't make the finished product safe for the insulators to use. When somebody asked him about better dust collectors, he said that profits wouldn't allow them now. He said that the amosite mines in South Africa were running out and that the rates were getting too high for the company to bear the cost. He also said it had been known since 1963 that asbestos could hurt us. At the same time, he said that amosite asbestos didn't cause trouble if you didn't smoke—that it wasn't proved medically that it could cause trouble. He had it written on the blackboard that seventeen per cent of the people who work with asbestos have lung cancer, but he told us that was another kind of asbestos. He claimed that if you didn't smoke you didn't have any more chance of catching cancer in the plant than you did walking out in the street. He also wrote a word I can't pronounce—meso-something."

"Mesothelioma?" I asked.

"That's it," Yandle replied. "Dr. Grant said for us not to worry about that, either. Especially if we didn't smoke. Cigarettes and asbestos don't go together is what he told us. Next thing we knew, they were sending us over to Dr. Hurst at the East Texas Chest Hospital for X-rays, lung-function tests, and blood tests."

I asked Yandle if he had ever been X-rayed before then, and

he said he had, on two occasions. "Both times, it was at the Medical and Surgical Clinic in Tyler," he told me. "They're the ones who handled all the accident work and medical stuff for Pittsburgh Corning. The first time I got X-rayed was in 1961, when I went to work there, and the next time was in the spring of 1969, not long after the Department of Labor people inspected the plant. According to Van Horne, all the X-rays they took of us in the spring of 1969 were good. We never saw them, of course, and wouldn't have known what to make of them if we had. I came across them, just by accident, last October, though, when the NIOSH boys were down here making their big inspection. I was taking them through the plant as the union representative, and I spotted a whole stack of something on a shelf high up in the men's room. So I climbed up there and took a look. It turned out to be our 1969 X-rays, but they were so old and cracked and covered with dust and dirt you couldn't have made out a thing."

I asked Yandle what he and his fellow workers were told about the X-rays and tests that Dr. Hurst had performed the previous August, and he said they were not told anything until a few weeks before the NIOSH inspection in October. "Early in October, two of us went over to the hospital to ask Dr. Hurst for our medical reports," he went on. "Dr. Hurst said that he had sent them to Van Horne and Dr. Grant. He explained that he was obligated to Pittsburgh Corning, because he had done the tests for them, but that, as a medical doctor, he was also obligated to us. He said we had no asbestosis on our diagnosis, and that if there were shadows on the X-rays they were probably not related to asbestosis. He said that as far as he could tell there were no health problems that were related to our work, but there were people who had emphysema and bronchitis, which he said was on account of smoking. When a NIOSH doctor examined me a few weeks later, though, he said he was sure I had a good dose of asbestosis. Also, I'd begun to think back on things and to wise up on my own. I remembered when Willie Hurtt had to quit a few years ago because he was spitting up blood. And

Robert Thomas, a neighbor of mine up in Hawkins, who hasn't been able to breathe good for a long time now. And Ed Land, with the same trouble. And Bill Morris, who was spitting up blood, too, and took himself over to the Gladewater Municipal Hospital in 1969, where they found spots on his lungs and made a biopsy. And a whole bunch of others I could name. So when Steve Wodka and I went over to the East Texas Chest Hospital in January to meet with Hurst, Grant, Farkos, Spiegel, and Lemen, and I heard that seven of us who'd worked in the plant for ten years had symptoms of asbestosis, I wasn't surprised. Wodka and I asked for a thorough medical reading on the health problems of each and every man in the plant, and after that we began to hear a different tune. Early in February, Van Horne put up a notice on the factory bulletin board scheduling appointments with Dr. Hurst for twenty-three of us, who, it turned out, had something wrong on our medical tests. I went over there on the ninth with Arthur Bearden, who'd worked in the plant for about sixteen years. I went in first, and Dr. Hurst read off a lot of numbers and said that after a recheck of my tests they'd made a new diagnosis. He told me that I had symptoms of early asbestosis but that my condition wasn't so bad as some of the others'. He said that a few more weeks in the plant wouldn't hurt but that after it shut down I shouldn't try to find work in a foundry or a welding shop, or anyplace where it's dusty, and he advised me to look for a truck-driving job. He told the exact same thing to Arthur Bearden. Naturally, one of the things I asked Dr. Hurst was what I could do for myself. And that's when it dawned on me that I was in real bad trouble, because he said that there was no medicine or cure for asbestosis, and that I had to keep coming back every two years for another examination to see if it was getting worse."

I asked Yandle how he felt these days, and he said not bad. "It's kind of hard to explain," he went on. "But if I exert myself real fast, I kind of just give out." Yandle added that in addition to Bearden and himself, Robert Thomas, Mitchell Walker, Tom

Belcher, and Harold Spencer had been told by Dr. Hurst that they had symptoms of early asbestosis, and had been advised not to work in dusty places. "There's a seventh guy with the same trouble, but he's a supervisor and he won't talk about it," Yandle said. "We hear that the company settled six months' pay on all the supervisors. These other fellows I told you about, though—they'll talk, and if you want I can introduce you to some of them."

I told Yandle I would like to meet them, and anyone else he could think of who could help to give me a clear picture of what had gone on in the Tyler plant.

"Well, then, you ought to talk to Frank Spencer," he said. "He's Harold's father, and the interesting thing about him is that he used to work at the old Union Asbestos & Rubber Company plant in McGregor, down by Waco. Old Frank's got an awful problem breathing these days, so he's almost always at home. Why don't I call him up now? I might as well call Ray Barron, too. Ray lives right close, and he's part of the maintenance crew they've kept on to clean up the plant."

Yandle telephoned Spencer and arranged for us to visit him at his house later in the evening. Then he called Barron and invited him to come to the motel. When he had hung up, I asked him to tell me what he knew about the burlap bags that asbestos was shipped to Tyler in and that were later sold to nurseries in the Dallas-Tyler area.

"That's a real laugh," he said. "I mean about them recalling thirty-five thousand of those bags from the local nurseries. You see, the amosite came in hundred-pound bags from South Africa to New Orleans and Houston, and then it was freighted over here on the Cotton Belt Route. Now, I happen to know that from 1963 to 1967 Pittsburgh Corning sold those bags to the Coastal Bag & Bagging Corporation, of Houston. Then, after 1967, they started selling them locally—mostly to the nurseries. Anyway, what I'm getting at is that the thirty-five thousand bags are just a drop in the bucket. Look at it this way. We used two hundred sacks of amosite a day in production, and we worked a five-day week. That makes a thousand bags a week, and fifty-two thousand bags a year. Is that right?"

"Right," I said.

"Okay," Yandle went on. "Fifty-two thousand bags a year for ten years gives you how much?"

"More than half a million," I replied.

"More than half a million," Yandle said quietly. "And you know what? They leaked like crazy—all of them. I know, because I used to help unload them from the freight cars. They leaked like crazy, and they were dusty as could be."

It was nearly seven when Ray Barron arrived—a handsome, strapping, dark-haired man of thirty-four, with a soft-spoken manner. He told me that he had worked at the Tyler plant for six years, first as a bag feeder, then in the building department, and now as a maintenance man. "I was a bag feeder for about six months," he said. "It was my job to empty the amosite out of the burlap sacks and into the feeding machines. I fed about seventy hundred-pound bags of asbestos on every eight-hour shift I worked. Trouble was, the asbestos wouldn't just pour out. It was packed in so tight we had to dig it out with our bare hands."

I asked Barron if he had ever worn a respirator when he was a bag feeder, and he shook his head. He had never worn one until August of 1971, he said, when the company suddenly made it mandatory for the entire work force to wear respirators. "Some of us were issued respirators a few years ago, but it was voluntary if we wanted to wear them, and since no one told us until last August that asbestos was dangerous to work with, we didn't," he said. "All of us in the maintenance crew are wearing 'em now, though. Except for Van Horne, who keeps saying that asbestos won't hurt you. He also claims that none of our medical problems are caused by asbestos. My own X-ray is supposed to be clear, and my other tests are okay, and I don't have finger clubbing, so they told me I was probably all right. What worries me is that word 'probably,' and whether I'll be all right five or ten years from now. When I think back, all I can remember is the dust. Why, most of the time I worked there you couldn't see from one end of the place to the other—especially when there was any sunlight coming through the windows."

Barron then said that for the past several weeks he and three other maintenance men had been dismantling machinery and equipment with acetylene torches and burying most of it in a dump next to the plant. "The little that's left is either being shipped up to Pittsburgh or being sold for scrap," he went on. "When we get through, the place'll be empty except for the surplus fiber that's still in the warehouse."

"How much of that is there?" I inquired.

"There must be about ten thousand bags," Barron answered. "I hear they've already sold it to some outfit up in Canada."

"Do you know who they are or where in Canada they are?"

"No," Barron said. "It's supposed to be a secret. But we'll be busy full time down there for at least six more weeks, so I might hear something. We've only just begun to bury the ventilation pipe, and we still have to finish cutting up the conveyor belts, carriages, ovens, and cyclone machines."

Barron left shortly after eight o'clock, and then Yandle and I drove out the loop highway—one of a series of roads that girdle Tyler—to the east side of town, where Frank Spencer and his wife live. We climbed the porch steps, and Yandle knocked on a screen door that opened into a small living room. Mr. and Mrs. Spencer were inside, sitting in rocking chairs and watching a quiz show on a large color television set. Spencer, a white-haired, ruddy-faced man, was wearing blue-and-white striped overalls. His wife, also white-haired, was wearing a plain calico dress. After Yandle had introduced us, I sat down on a small couch and asked Spencer to tell me about his experiences working with asbestos.

"Well, I worked with asbestos for twenty years or so," he said. "I started in 1948 or 1949 in the old plant down in McGregor, where we used to live. The factory was in the Bluebonnet

Ordnance Plant area, where they'd made bombs for the Air Force during World War II, and the government began reactivating the place in 1952, so Union Asbestos & Rubber moved out and came up here in '54. There was about sixty of us on the work force at McGregor, but there was a big turnover. Most of us were farmers, and there was a lot of young farm boys who came and went. Harold, my own son, worked there, too. In fact, I signed a paper saying he was eighteen, but he was really only seventeen. That's so he could get the job."

I asked Spencer if he had ever worn a respirator at McGregor, and he gave me a look of amazement. "Wore a *res*pirator!" he exclaimed. "Why, I wore one all the time down there. They never said a word to us about asbestos being dangerous, but they made us wear our respirators. Actually, it wasn't the company that did that. It was the insurance underwriters who said we had to wear 'em. I guess they must have known that that dust couldn't be doin' us much good, eh? But, goodness' sakes, insurance underwriters or not, you would have *had* to wear a respirator down there—you couldn't have breathed otherwise. Most of the time, I couldn't hardly see the man working next to me!"

"Do you ever get back there?" I inquired.

"We used to go back all the time to see our friends and relatives," Spencer replied. "But, one by one, just about all the men who worked with me at that place have died. Mostly, they just seemed to stop breathing."

"Some of them died of cancer, too," Mrs. Spencer added, without taking her eyes from the television set. "Chest cancer, the womenfolk told me."

"I'll be sixty-six years old in May," Spencer said. "And I swear I'm just about the onliest one I know that's left alive from that McGregor plant. Except for my son, of course, and he's already got trouble with his lungs. I've had trouble with mine for years. I finally had to quit the Tyler plant in 1968. They disabled me. Just couldn't hardly breathe no more. Now I'm out of breath all the time. Can't do nothing. Can't walk any distance at all. When I quit, they started giving me some retirement payments—eighteen dollars a month. I should've done what Ed Land did. He quit around about 1962. He couldn't breath, either, because a lung collapsed on him. I heard he was going to sue, but they settled with him."

I asked Spencer if he was receiving any medical attention, and he smiled for the first time. "I've been going to these local doctors for my lungs for years," he said. "I spit up an awful lot, and I'm always out of breath, and I've got a funny-sounding cough. I guess I've been to just about every last doctor in that Medical and Surgical Clinic. All of 'em have told me I had something wrong on my X-rays. Some of 'em have told me I got emphysema, and some of 'em say it's asthma, and at one time they even thought I had cancer, but not one of 'em ever said it was on account of asbestos. Well, they're a bunch of quacks. I know what I got now. I got what Ed Land got. I got what Harold's got. And all the others. I got that dust disease. That dust has ate us up."

As Yandle and I left a few minutes later, Spencer rose from his rocking chair, saw us out the door, and stood on the porch and bid us goodbye. As we went down the steps, I turned to thank him, and saw him leaning forward, hands on his knees, rocking his head up and down. He was gasping for air.

At eight o'clock the next morning, Yandle came by the motel in his battered blue pickup. We had a cup of coffee in the dining room and then drove out to Arthur Bearden's house, on the east side of town. Bearden's place was a faded-green frame dwelling that sat on cinder blocks, like most of the houses in the area, and in the front yard there was a twenty-five-foot-high, symmetrically rounded, and wonderfully festive holly tree, so full of shiny green leaves and bright-red berries it looked as if it had been hung by hand.

Bearden came to the door, ushered us inside, and immediately offered to brew us some coffee. He is a grave-faced, soft-spoken man in his middle forties, and when Yandle told me he was famous for making strong coffee he responded with a broad grin of delight. Then, seating me at a table in the dining room, he disappeared into the kitchen. Yandle went to get Harold Spencer, who lived in a trailer next door.

When Bearden returned from the kitchen, he sat down across from me and told me that he had gone to work at the Tyler plant in March of 1955, a few months after it began operating. "I worked there until 1960," he said. "Then I was laid off for a year, but I went back in April of 1961 and stayed until the plant closed down last month. So, all in all, I figure I worked with asbestos about sixteen years. During that time, I guess I worked in most areas of the production department. I never wore a respirator until last August, and I never saw anyone else wearing one except for a few fellows on the feeder machines and in the saw room back in 1969. Those were probably the dustiest areas in the plant, but the whole place was so dusty it's hard to tell for sure. Most days, you couldn't see from the middle of the building to either end."

At this point, Yandle returned with Harold Spencer, a man in his late thirties, who wears a grin that masks a diffident manner and an air of fatalism. When I told him I had met his father and mother the night before, he simply nodded and grinned a little wider, and sat down at the table. I asked Spencer how long he had worked with asbestos, and he thought for a few moments and said he guessed for about seventeen years. "I worked at the Union Asbestos & Rubber plant in McGregor from '51 to '53," he said. "Then I worked at Tyler from '55 to '57, and then I came back in 1960 and worked steady from then on until the plant closed down last month. Nobody ever told me the stuff would hurt us, and I never wore a respirator until last August, when everybody had to start wearing them. Back around last Thanksgiving time, I got called over to the East Texas Chest Hospital by a Dr. Seaman, who told me I had a spot on my left lung. He didn't say what it was, but he offered to treat me as an outpatient. He said I could keep on working at the plant that way. I went to him for about two months. During December and

January, I went once a week, on Tuesday morning. They took X-ray pictures of me every time. Most mornings, they just took a couple, but one morning they took eleven—one standing up and ten flat on my back. Another time, they swabbed out my mouth with anesthetic, put in some kind of a coating, and looked down into my chest with some kind of a tube. Afterward, Dr. Seaman told me my left lung wasn't functioning properly. I asked him if it was on account of the asbestos, but he said he couldn't say. All he could say was that there wasn't any surgery for it. Then, on February 9th, I got sent over there to see Dr. Hurst. He said I had what looked to him like asbestos scar tissue in my lungs. He told me not to work in dusty conditions any more, and he said to come back and see him in two years. By that time, I was real worried and didn't know what to think. What worried me most of all was what a doctor from NIOSH told me back in October. He had listened to me breathe, and he had looked real careful at my fingers, and he had told me I had a case of asbestosis."

"That's what we all got," said Bearden, looking around the table at his friends and then at me. "We just better hope we don't get cancer. You know, some people have already died of cancer who worked there. Mr. McMillan—he was plant manager before Van Horne—he died about a year and a half ago of some kind of cancer in the chest. And Marvin King, one of the supervisors—he took sick back in 1958, I think it was, and died of lung cancer. And Chester Hickman, who'd been night watchman since the place opened—he died of cancer in June of 1971."

Bearden went out to the kitchen and brought back coffee for everyone. It turned out to be as strong as any I had ever tasted, and when I told him so he grinned with delight again. "Spencer here was the only one of us that got treated as an outpatient," he said. "Back in October, Dr. Spiegel told me I had a bad sound in my lungs, but I didn't see Dr. Hurst until I went over to the East Texas Chest Hospital with Herman here in February. Dr. Hurst told me I had early stages of asbestosis with rales in my lungs, and that I shouldn't try to find a job anyplace there was dust.

The thing that bothers me is I should've realized something was wrong back in 1968. I was off from work then with a bladder infection and got sent over by the company to see Dr. Smyth, at the Medical and Surgical Clinic. He gave me a physical examination and listened to my chest and took some X-rays. A few hours later, he called me back and listened to my chest again, but he didn't tell me anything."

"Funny thing," Spencer murmured. "Back in 1969, I got my first X-ray at the Medical and Surgical Clinic, and a Dr. Marshall over there told me I had a spot on my lung. He said he'd be in touch with me, but after I waited six months without hearing from him I got worried and went back and saw a Dr. Whetsell there. Dr. Whetsell gave me a checkup, and when he was through I told him I thought that maybe I ought to change jobs, but he said he couldn't guarantee I'd be hired anywhere else, because I might not be able to pass a physical."

While Spencer was talking, we were joined by Willie Hurtt, whom Yandle had invited over. Hurtt is a gaunt man in his late fifties, with sunken cheeks and a shy manner. He told me that he had started work at the Tyler plant in January of 1956, and had stayed until November of 1969, when he quit. "I worked mostly on the feeding machines and the ovens, and I hardly ever wore a respirator the whole time," he said. "I remember, though, that around the end of May of 1969 some fellows from the union committee went in and asked Van Horne how come a few of the boys were suddenly wearing masks. Van Horne told them it was on account of the dust levels were high in the saw room and the feeding-machine area. So they asked him if the dust would hurt us, and Van Horne just said that even if it did there wasn't any way anybody could be sure of it. Of course, by that time I was beginning to get some idea of my own that the stuff wasn't good to breathe. I'd been coughing bad for nearly three years, and I'd been to a lot of doctors. I went to see Dr. Thomas and Dr. Smyth at the Medical and Surgical Clinic, and I went to Dr. Knight at the Glenwood General Hospital, in Tyler. Dr. Smyth told me I ought not to work in dust, but he never said I'd be disabled. Dr. Knight said I had emphysema in both lungs. He kept me at the Glenwood General Hospital for a week after I quit the plant. He took a lot of X-rays and asked me where I worked, but he didn't tell me anything about the X-rays and he didn't say anything about asbestos. Now I'm not worth a hoot to do anything. If I walk just a little piece, I can't even breathe."

"Same with me," said Bearden. "I hoed a row in the garden the other day and I was out of breath entirely."

"The worst thing of all is we can't pass a physical to get another job," Spencer said.

Bearden nodded soberly. "I got an application in at a plant that manufactures air conditioners," he told me. "But I'm a-scared to death of the physical. I'll take it, but I'm afraid of it, 'cause I know I won't pass it if they take an X-ray or listen to my lungs." For a moment, Bearden was silent, looking around the table at his friends. "You know," he said reflectively, "not everybody was as dumb as we were. I remember a new man walking in a few years ago. It was in 1970, I think. They had told me to explain the feeders to him, and it was nighttime and the lights were on and the dust was so thick it was fogging the lights. This fellow—I'll never forget him. He took a look at the feeders, and he walked partway down the plant floor. Then he turned and walked back past me and gave his head a shake. And then he went right on out the door, without a word to anybody, and he never came back."

It was about eleven o'clock when Yandle and I returned to the motel. I had an appointment to see Van Horne in an hour, and Yandle had to do some work around his house. We arranged to meet there late in the afternoon. Half an hour later, I drove east to U.S. Highway 271, where I turned north at a huge foundry that was belching smoke into the sky. About eight miles down the road, I passed a large, modern brick building surrounded by expansive lawns. A sign identified it as "East Texas Chest Hospital." A few hundred yards past the hospital, I turned west onto State Highway 155, where there was a sign pointing the way to the Owentown Industrial District. At the bottom of a hill, about a mile northwest of the hospital, I came to a flat valley filled with sheds, prefabricated buildings, and some wooden warehouses that had once been part of Camp Fannin, a World War II training center and POW camp. The whole place was crisscrossed by spurs of the St. Louis Southwestern Railway. After winding through the area for five or ten minutes, I came upon a pair of long, low sheds with peeling wooden façades and corrugated roofs, and saw a sign affixed to the side of one of them which read "All Persons Entering This Area *Must* Wear Approved Respirator Protection." It was a new white sign with bright-red lettering, and against the shabby background it shone like a beacon. At one end of the sheds was a green cinder-block office building with a sign reading "Pittsburgh Corning Corporation."

I parked the car out front, went inside an anteroom, and gave my name to a receptionist who announced my arrival on an intercom. A minute later, Van Horne came out, shook my hand, and ushered me into his office, where we sat down near his desk. He was a gray-haired, self-assured man in his middle forties, and he was wearing horn-rimmed glasses and a mustard-colored sports jacket. When I told him that I wanted to talk with him about the plant, he said he had been in touch with the Pittsburgh office an hour before and had received certain orders. "I have been instructed to read a statement to you," Van Horne continued. "It is an authorized statement, and after I am done reading it I will have no further comment." Having said this, he got to his feet, picked up a sheet of paper from the top of his desk, adjusted his glasses, and, in the firm, measured tone of a schoolmaster explaining the rules to a new and possibly recalcitrant student, began reading. "Pittsburgh Corning, on its own initiative, has closed the Tyler, Texas, plant and has ceased asbestos production entirely," the statement said. "We have no plans for further use of the Tyler plant and are making every effort to insure safe disposal of any remaining waste material. We sincerely regret the adverse economic impact that the plant closing might have on our employees and their families. Since 1967, we have made efforts and investments to reduce dust levels at the plant and have kept government agencies advised. It has become economically impossible to reach the increasing rigid dust levels established by the federal government. Each of the employees at the plant have been given complete physical examinations and, in cooperation with local officers of the Oil, Chemical, and Atomic Workers Union, we have established a job-counseling service to aid men in locating other work. There is no substantiation for the claim that burlap bags from the plant are a health hazard. Despite this, we have recently bought from nurserymen in the Dallas-Tyler area some thirty-six thousand bags, which may have come from our plant."

When Van Horne finished reading, there was a moment of constrained silence; then I got to my feet, thanked him for giving me his time, and asked him if I could have a copy of his statement. He rang for the receptionist, who came in, took the statement, went out, and returned with a Xerox copy. Van Horne handed me the Xerox, shook my hand, accompanied me to the front door, and wished me a pleasant trip back to New York City. I told him I was planning to stick around town for a while. Then I got into the car and drove away.

When I got back to the motel, I went to my room to make a few phone calls. While I was looking up some numbers in the local directory, I happened upon a description of the town, and learned that it had been established in the early eighteen-forties and named for President John Tyler, who had advocated the annexation of Texas to the United States. According to the phone book, Tyler had been "a small agricultural town with an aristocratic background" until 1930, when the great East Texas oil field was discovered. At that point, the phone directory said, "Tyler became the office headquarters for two hundred and sixty-seven independent and major oil producers, operators, and refiners working in the fields surrounding the city." Beneath the description was a box with the heading "Some Interesting Facts

About Tyler." One of the interesting facts was that Tyler has a twenty-eight-acre municipal garden containing twenty-five thousand rosebushes in three hundred and fifty varieties. Another interesting fact was that "more than half of the field-grown rosebushes supplied to the entire nation come from within a fifty-mile radius of Tyler." The most interesting fact about Tyler I saw, however, was that it is the medical center of East Texas, with an expanding complex that includes the Medical Center Hospital, the Mother Frances Hospital, two other general hospitals, four private clinics, and, on the outskirts, the East Texas Chest Hospital, which has eight hundred and twenty-five beds.

After lunch, I drove out to the East Texas Chest Hospital to keep my appointment with Dr. Hurst. It was a windy afternoon, with clouds scudding across the sky, and on a flagpole in front of the main entrance to the hospital the American and Lone Star flags were flapping to the accompaniment of rattling grommets. A lot of Texas redbud trees on the grounds were just beginning to bloom, and they cast a sheen of purple against the well-trimmed green lawns. Some distance off, ramshackle wooden barracks with sagging roofs, peeling clapboard walls, and verandas evolving into debris indicated that, like the nearby Owentown Industrial District, the site had once been part of Camp Fannin.

After spending a few minutes in a waiting room, I was directed to the office of Dr. Hurst, whom I found to be a tall, thin man in his late thirties, with dark hair, an earnest manner, and a mild way of speaking. Born in Minas Gerais, in Brazil, where his father was a missionary, Dr. Hurst received his training at the University of Texas Southwestern Medical School, in Dallas; he then became a specialist in pulmonary disease, and for several years was a member of the staff of the Chest Division of Woodlawn Hospital, in Dallas. In 1964, he was appointed clinical director of the East Texas Chest Hospital, and in the autumn of 1970 he became superintendent of the institution. He told me that the hospital had been founded in

1949 as a tuberculosis hospital, and that its major work had always been in the field of tuberculosis treatment. "Since 1967, however, we've been treating patients afflicted with other kinds of pulmonary disease," he said. "The fact is that when Dr. Grant, as medical consultant to Pittsburgh Corning, approached me late in 1966 about conducting a medical survey of the workers in the asbestos factory here, it promised to be a real departure for us."

"What reasons did Dr. Grant give you for wanting the study performed?" I asked.

"He told me that the men were working with amosite asbestos, and that the company wanted to know if they were encountering any medical problems as a result of their exposure."

I remarked on the irony of the factory's proximity to a hospital specializing in pulmonary disease, and asked Dr. Hurst if he had known of the plant's existence before Dr. Grant's visit.

Dr. Hurst shook his head somberly. "No," he said. "I'd never heard of the place."

"Did Dr. Grant give you any indication that he suspected the men might be getting sick because of their exposure to asbestos?"

"No, he did not," Dr. Hurst replied. "Obviously, he knew that exposure to asbestos could cause pulmonary fibrosis. He simply told me that the company had been using the Medical and Surgical Clinic, and that it now wanted us to undertake a comprehensive medical survey of its workers."

I asked Dr. Hurst if he had ever been consulted by any doctor at the Medical and Surgical Clinic about any worker at the Pittsburgh Corning plant who had had to stop work because of pulmonary insufficiency.

"No," he said. "Never. I only examined men sent me by the company, who were working and were apparently well."

I told Dr. Hurst that several of the asbestos workers I had talked with had said that doctors at the clinic had told them they were afflicted with emphysema or bronchitis, and that I had not yet met anyone who had been told by any doctor at the clinic that he was afflicted with asbestosis. I asked Dr. Hurst if this seemed strange to him.

For a moment, he was silent. Then he said, "You know, there's an awful lot of ignorance about occupational disease. The average doctor tends to minimize the danger of it. Some doctors tend to blame everything on cigarettes."

I asked Dr. Hurst if he had learned at any time that Dr. Lewis Cralley, of the Bureau of Occupational Safety and Health, in Cincinnati, had not performed the medical survey of the Tyler workers he had agreed to back in 1967.

"No, I did not," he answered. "I assumed that the study had been conducted, as Dr. Grant had indicated it would be."

"What reasons did Dr. Grant give when he called you last August and once again proposed that you examine and test the workers?" I inquired.

"The same reasons he gave nearly five years before," Dr. Hurst replied.

"Did he tell you that Pittsburgh Corning had filed a request with the Department of Labor for a variance from occupational-health regulations, and that the request for a variance stated that the company had already expanded its medical-examination program?"

"No," said Dr. Hurst.

When I asked Dr. Hurst to describe the tests that were performed on the Tyler workers at the East Texas Chest Hospital in August of 1971, he told me that they included ventilatory studies, to determine how fast air goes into and out of the lungs; lung-volume studies, to show how much air the lungs can hold; blood-gas studies, to indicate how efficiently oxygen is getting into the bloodstream; diffusion studies, to indicate what happens to oxygen and carbon dioxide in the lungs; and chest X-rays.

I then asked Dr. Hurst when he first suspected that some of the workers were afflicted with asbestosis. "Well, that took a while, because we weren't able to pull all our data together until

November," he said. "By then, we knew there was evidence of fibrosis, or scarring of the lungs, on some of the X-rays. Some of the men also had dyspnea, or shortness of breath, and there were decreases in diffusion capacity as well. In fact, all but one of the seven men in real trouble had low diffusion. The trouble was, we had made a mistake in our formula for the diffusion studies, and didn't discover it until December. That's why it took us so long to come up with a positive diagnosis of asbestosis."

I then asked Dr. Hurst if he or anyone else was planning to perform followup medical studies of the Tyler workers or their families, or of former workers or their families.

"I've talked about that with Dr. Johnson, at NIOSH, with Dr. Norman Dyer, of the Environmental Protection Agency, in Dallas, and with my superiors in the Texas State Department of Health," he said. "I really can't say what's going to happen, however. I suppose it's a question of where the money will come from."

Next, I asked Dr. Hurst if the East Texas Chest Hospital had ever before examined workers on a consultant basis with a company.

"Not to my knowledge," he replied.

"Would you do it again?"

"I don't know," Dr. Hurst said slowly. "The studies we performed for Pittsburgh Corning were good studies, and once we rectified the error in our diffusion tests they were thorough. But whether we should have got involved with the company in the first place is something I'm not sure about."

It was about three-thirty when I left the hospital, so I took the shortest route to Yandle's house, driving north on State Highway 155, past the Owentown Industrial District. I continued, through Winona, to the town of Big Sandy, where I turned west on U.S. Highway 80, toward Hawkins. At one point, I crossed a bridge over a muddy, meandering stream, and then the road passed through mile after mile of pecan groves, where snow-white cattle grazed among thousands of frail black trees, whose delicate branches formed an intricate tracery against the sky. The center of Hawkins consists of four corners, where U.S. Highway 80 is bisected by State Highway 14. A sign there says that the town has a population of nine hundred and seventy-seven. I turned north on State Highway 14 and, within a mile or so, arrived at Yandle's place—a small white frame house—and found Yandle standing in the driveway.

Yandle and I spent the rest of the afternoon hunting up men who had worked in the Tyler plant. A mile down the road, we paid our first call, on Dale Peek, a lean, good-looking fellow in his late twenties, whom we found in the yard behind his house, bending over the engine of a car. Peek was wearing a navy-blue wool watch cap, a shirt with the sleeves cut out, and Levi's. After we shook hands, he stuffed his hands into the hip pockets of his Levi's, sat down on a fender of his car, and told me that he had worked at the Tyler plant for five years. "Like most of the other boys, I had a whole bunch of jobs," he said. "I was a feeder, a builder, a saw-room laborer, and a scale man. Scale men are the ones who weigh the boxes carrying the finished product. I finally quit the place in July of 1969, on account of the respirators they had started making some of us wear. I couldn't breathe through the damn thing. I just couldn't get any air into me, so I quit. Now I work making corrugated boxes at the Continental Can plant. That's in the Owentown Industrial District, too. The spring before I quit, I'd been elected chairman of the union committee, and I went in to Van Horne to ask him about some articles that some of the fellows were reading that said asbestos was bad for you. Van Horne passed it off with a shrug. He said that as far as he knew the doctors had no way in the world of knowing for sure that the stuff caused you to get sick."

With Peek accompanying us, Yandle and I drove west about fifteen miles to the town of Mineola, to see Tom Belcher, one of the seven men whose symptoms had been diagnosed as those of asbestosis, and who had been advised by Dr. Hurst not to work in a dusty environment. On the way, Yandle told me that

Belcher had worked in the plant for sixteen years, and that he had started as a feeder operator.

"Him and Willie Hurtt," Peek said. "They're real old-timers at that place. I used to see them shaking down the dust collectors. Why, the stuff was so thick on them you couldn't hardly make out their hair or faces. I remember Tom sometimes wore a mask. It wasn't a real mask, though. It was one of those gauze ones, like you see them wearing in hospitals on the TV."

Belcher was a florid, heavyset man in his late forties, and as Yandle, Peek, and I stood talking with him in the front yard it was obvious that he was having difficulty breathing. "No one ever told me the damn stuff was dangerous," he said bitterly. "About two years ago, though, I started to feel different in my chest. It felt tired all the time, and it hurt a lot. It's been that way ever since, and these days I cough so much in the morning I can't hardly stand it."

A few moments later, Belcher's wife called me into the house and showed me a copy of Dr. Hurst's diagnostic report. According to the report, Belcher had a reduction of forced vital capacity, which is the maximum amount of air he can take into his lungs; a twenty-five-per-cent reduction in total lung capacity, which is the amount of air his lungs can hold; and a slight reduction in the amount of oxygen that gets into his bloodstream. In addition, his X-ray films showed an increase in bronchovascular markings on the lower left-lung field. Mrs. Belcher said that her husband's chest was giving him a great deal of pain, that he was worried about it, and that she was worried about his worrying, because in addition to his pulmonary problems he had high blood pressure.

When we left the Belchers, Yandle, Peek, and I drove to a place about seven miles north of Hawkins, where J. C. Yandle, Herman's brother, lived, with his wife and family, and worked as a tenant rancher. He had a typical small Texas spread, with a long blacktop drive leading up to a native-rock farmhouse that was surrounded by pastures separated by well-kept fences. We found J. C. Yandle in a field adjacent to the house, fixing a cultivator rig. He was a tall, thin, lean-faced man in his middle forties; he was wearing Levi's and a denim jacket, and he looked as if he had stepped right out of a scene in *The Last Picture Show*. After Herman introduced us, J. C. leaned against a corral fence and, speaking very deliberately, told me that he had gone to work at the Tyler plant in November of 1961. "I worked there steady until the place shut down last month," he said. "During that whole time, no one ever told me or anyone else I know that asbestos could harm you. Why, I can remember some of our supervisors saying it not only wouldn't hurt you but was *good* for you! They even used to tell us you could *eat* it. They were saying that right up till August, when all of a sudden we have to wear respirators. First thing I heard about my X-rays was on the ninth of February, when Dr. Hurst called me in and said they didn't show up anything wrong. He told me I was breathing in better than I was breathing out, though. I asked him why. He said some people do that. He said maybe I was smoking too much. Funny thing is, I roll my own and I don't smoke an awful lot. About half a pack a day, maybe. But I am sure awful short-winded, and for the last few years it's been getting worse and worse. I been coughing up a terrible storm in the mornings, too."

After we said goodbye to J. C. Yandle, his brother and Peek and I drove toward Hawkins on State Highway 14 and, about a mile before Herman Yandle's place, stopped at a frame house with brown asbestos siding where Ray Hicks, a former Tyler worker, lived with his mother. Hicks wasn't at home, so we drove a little farther along the road, to a house where Hubert Thomas, another of the Tyler workers, was visiting some friends. As we were getting out of the car, Hubert's brother Robert drove up, with his wife, and parked just in front of us. Robert Thomas was fifty years old, he had worked at the Tyler plant for ten years and three months, and he was one of the seven men whose symptoms had been diagnosed as those of asbestosis. "At the end, just before they closed the place down, I was working in the finishing department," he told me. "It was my job to push cartons of finished insulation down along the factory floor. Trouble was, I just couldn't push the damn things without stopping every hundred feet to get my breath. My chest's been hurting something awful lately, and I've been short of wind for the past five or six years. I also cough a lot these days. When I got called in by Dr. Hurst last month, he told me that he and the other doctors at the hospital had decided I probably had emphysema or bronchitis. He also said I could have symptoms of a disease caused by all the dust I'd breathed over at the factory. He advised me not to accept any work near dust, like at one of the local pipe or foundry outfits, and he told me to quit smoking. He said that I should have another X-ray soon and that the ones they'd already taken showed something wrong. I think he said it was some enlarged tissues down there. When he told me that, I remembered something funny that happened back in the last part of 1970. I'd hurt my back, and the company had sent me to the Medical and Surgical Clinic, which shipped me over to the Mother Frances Hospital for some X-rays. After they took the X-rays, I was lying in my bed when a nurse came into the room with a clipboard. She looked at the clipboard for a minute, and then she asked me if I was there for my back or for my lungs. I told her I was there for my back, and she shook her head and said that that was odd, because there sure was something wrong with my lungs."

At that point, I said goodbye to Robert Thomas, and walked up a driveway toward the house. The driveway crossed a deep culvert by the roadside, and on the other side of the culvert was a large tank full of channel catfish, which are a delicacy in that part of Texas. Two children, armed with long-handled dip nets, were trying to catch a pair of them for a customer who had stopped off to buy his supper. The catfish were surprisingly agile, and the children were having a lot of fun but not much luck. When Hubert Thomas strolled over to see if they needed a hand, I introduced myself to him, and we stood talking for some time beside the tank. He told me that he had gone to work at the

Tyler plant in October of 1961, and he had stayed on until the factory was shut down. "We're all kind of worried about what's going to happen to us," he said quietly. "Especially those of us that's got a lot of years in that place. Nobody's told me anything about my medical examination yet, which may be a good sign, but I can tell you something for sure. I'm so short-winded these days it's ridiculous. I can't even walk from the tank here to my house without getting out of breath. Why, I not only have to stop a couple of times to get it back but sometimes I have to sit right down on the ground!"

I asked Thomas if he had ever worn a respirator and if anyone had ever told him before Dr. Grant's visit that asbestos was dangerous to work with.

He shook his head. "I never wore a respirator until last summer," he replied. "And nobody ever said to me that the stuff could hurt you. No—come to think of it, that ain't right. There *was* a man who told me asbestos was dangerous. That was Mr. McMillan, who was plant manager before Van Horne. Way back, when I first went to work there, he told me the stuff could be harmful. He never said to wear a mask, though, so I didn't. I got to respect that man, because nobody else from Pittsburgh Corning was ever that honest with me. But it's funny, ain't it? I mean, Mr. McMillan knew it was dangerous, and he died of cancer himself not long ago, but he never said to wear a mask. So I didn't. None of us did."

It was growing dark, and when Yandle, Peek, and I left, we drove south on State Highway 14 and stopped again at Ray Hicks's place. His mother came to the door and told Yandle that Ray had come and gone since we were there before but that she was pretty sure we could find him at the service station down by the four corners in Hawkins. A few minutes later, Yandle spotted Hicks's car at the service station, and when I drew up beside it he climbed out to get him. Hicks, a tall, thin, handsome, pale man in his early thirties, got into the back seat with Peek and Yandle got in front with me, and we spent a few minutes chatting. Hicks told me that he had gone to work at the Tyler plant on April 30, 1962, and had worked there until the factory was shut down. "That makes nine years and ten months, and I never wore a respirator except for the last six months I was on the job," he said. "I never knew it was dangerous to work with asbestos until last summer, when Dr. Grant came down. He told us it was hazardous, but he said you didn't get any symptoms in less than ten or fifteen years."

I asked Hicks how he felt, and he said that he felt pretty good—that, unlike a lot of his friends from the plant, he wasn't experiencing any short-windedness. "Dr. Hurst took X-rays of me in August of 1971, but I never heard from him, so I guess that means I'm okay," he added.

"Hey, why don't you tell us what Van Horne told you when he called you in last month," Yandle said, with a smile.

"Oh, that," Hicks said with a sheepish grin. "Why don't you guys lay off that?" Hicks then shook his head and smiled ruefully at me. "Well, Van Horne called me into his office just before the place shut down," he said. "He told me a report on my X-rays indicated that there was a little spot on the lower lobe of my left lung, and he said I'd better stop smoking. I told him I'd never smoked in my life—except cigars. So he asked me if I drank a lot of milk."

When Hicks said this, Yandle began to chuckle. Then, joined by Peek, who had broken into a guffaw, both men leaned back in their seats and gave way to gales of laughter.

"And what did you tell him?" asked Yandle, making a vain attempt to stifle his laughter.

"I told him I drank a little milk now and then," Hicks said. "That's the truth, too. I don't drink all that much milk."

"A little bit?" said Peek, who was suddenly overtaken by another paroxysm of mirth. "Just a little bit?"

"And what did Van Horne say when you told him that?" Yandle cried, still laughing.

Hicks glanced at me, but he had begun laughing himself now,

and suddenly he doubled over as the mirth bubbled out of him. "Van Horne told me I had a calcium deposit in my lungs, and said I must be drinking too much milk."

At this point, Hicks broke down completely, and I began to laugh myself. For some moments, the four of us sat rocking with laughter, unable to speak; then Hicks straightened up and managed to get partial control of himself.

"I walked out of there real mad," he said. "But when I told my friends back on the floor, all they did was laugh, and they've been ribbing me ever since."

"But, Ray, we was worried about you!" Yandle protested. "We was worried about you on account of all that milk you've been drinking."

This comment started everybody laughing again, and we were still laughing when we climbed out of the car, a few minutes later. I thanked Yandle for all his help, shook hands with each of them, and bade them farewell and good luck. Then I got back into the car and started for Tyler. As I drove away, I caught a glimpse of Yandle, Peek, and Hicks in the rearview mirror. There they were, in front of the filling station, with their arms around each other's shoulders and their heads thrown back—three Texans standing beneath the stars of the immense night sky and laughing to beat the band.

In the morning, I ate breakfast, paid my bill, and checked out of the motel. I was planning to drop by the Medical and Surgical Clinic to ask the doctors there why none of the men who had come to them with pulmonary problems had been told that the trouble might be the result of the place they were working and what they were inhaling on the job. Some of the men I had spoken with had been told they had bronchitis, and others had been told they had emphysema, but none of the doctors at the clinic, as far as I could make out, had ever told them they were suffering from asbestosis, which is, of course, what was afflicting many of them, and was the reason they were experiencing such difficulty in breathing. However, after sitting behind the wheel of my car for a few moments I decided not to go to the Medical and Surgical Clinic. What good would it do to ask a bunch of local doctors how they could have failed to diagnose a lung disease that had been well known and well defined in medical circles for more than fifty years, when men such as Dr. Grant, medical consultant to Pittsburgh Corning, medical director of PPG Industries, president of the American College of Preventive Medicine, and the holder of several other high posts in distinguished industrial health organizations, and Dr. Cralley, director of the Division of Epidemiology and Special Services of the United States Public Health Service's Bureau of Occupational Safety and Health, not only had known for years that workers at the Tyler plant were being subjected to levels of asbestos dust that could cause asbestosis but had neglected to give them medical examinations that would have ascertained this beyond a doubt? All of a sudden, I felt I had asked enough questions for the time being, and I was ready to accept Dr. Hurst's explanation that most doctors are ignorant of occupational disease and prone to blame cigarettes for pulmonary trouble. In any case, what sort of judgment could one make about the doctors at the Medical and Surgical Clinic, who, after all, had had a close working arrangement with Pittsburgh Corning over many years?

Having decided not to go to the Medical and Surgical Clinic, I thought I would return to Dallas and fly home. I wanted to take a last look at the factory, however, so I drove out past the East Texas Chest Hospital and turned north on State Highway 155. It was about 10 a.m. when I came to the Owentown Industrial District, and as I wound my way through the area, which is about a mile long and half a mile wide, I saw several hundred cars parked outside the factories, sheds, and warehouses of the other industries there. Among those industries were the Continental Can Company, where Dale Peek now made corrugated boxes; the National Homes Corporation, makers of house trailers and prefabricated housing units; the Ty-Tex Rose Nursery; Levingston-Armadillo, Inc.; National Casein; and the

Texas Tubular Products Company, next door to the Pittsburgh Corning plant, which Ray Barron and the other men of the maintenance crew were dismantling and burying. All in all, there must have been close to fifteen hundred men and women working in that industrial park, and, remembering the case of the proprietor of the junkyard adjacent to the old asbestos plant in Paterson, New Jersey, who had died of mesothelioma, and the documented cases of dozens of other mesothelioma victims whose only exposure to asbestos was simply that they had lived or worked in the vicinity of asbestos factories, I could not help wondering what would happen to these Owentown workers. Then I remembered the case of the daughter of the engineer who developed the product that had been manufactured in the Tyler plant for seventeen years—she had died of mesothelioma, though her only known exposure to asbestos had occurred when she played with samples of asbestos products her father brought home from time to time—and the cases of dozens and dozens of other mesothelioma victims whose only known exposure to asbestos was that as the wives of asbestos workers they had washed their husbands' work clothes, or as children and relatives of asbestos workers they had simply lived in the same house, and I found myself wondering about the fate of the families of the eight hundred and ninety-five men who had been employed at the factory during its seventeen years of operation. At that point, the Owentown Industrial District became too depressing for me, and I drove away. A few miles down the road, I came to Interstate 20, and turned east toward Dallas. It was a warm, sunny morning; buds were bursting everywhere, and the green sheen I had noticed on the way to Tyler two days before had become palpably greener. Overhead, the buzzards were out in force.

PART THREE

Some Conflicts of Interest

Shortly after I returned to New York, I arranged to fly to Cincinnati and spend a day with Dr. William Johnson and Dr. Joseph Wagoner, of the NIOSH Division of Field Studies and Clinical Investigations. I also telephoned Dr. Lee Grant at his office at PPG Industries, in Pittsburgh, and asked him if he could spare an hour or so to talk with me about the Tyler plant. Dr. Grant was extremely cordial, but he declined to give me an interview unless I first obtained the permission of James Bierer, the president of Pittsburgh Corning. I then called Bierer, and he, too, was very cordial, but somewhat hesitant regarding my request. He said that he would have to look into the matter before giving me permission to talk with Dr. Grant. "I'll get back to you as soon as possible," he said.

On Monday, March 13, 1972, I took a morning flight to Cincinnati, and arrived at the offices of the Division of Field Studies and Clinical Investigations shortly before noon. Dr. Johnson turned out to be a tall, pale, bespectacled man of thirty-one, with a quiet way of speaking and a serious demeanor. His boss, Dr. Wagoner, was a boyish-looking blue-eyed man of thirty-six; like Johnson, he is extremely soft-spoken, but his manner is more intense. I had a lot of questions for them about the survey they had conducted at the Tyler plant, in October of 1971, and by the time we had finished with these we were in the middle of lunch at a nearby restaurant. At that point, I told them something about my recent trip to Tyler, and how I had met several men who had become ill and stopped working in the plant even before it was shut down. When I finished giving them my impressions of these men, Dr. Johnson put down his fork and shook his head.

"As you know, Dr. Irving Selikoff and Dr. E. Cuyler Hammond have conducted a study of the mortality experience of nine hundred and thirty-three men who worked between 1941 and 1945 at the Union Asbestos & Rubber Company's plant in Paterson, New Jersey, which was the predecessor factory to the one in Tyler," he said. "Because of their findings, we're awfully depressed about the future of many of the eight hundred and ninety-five men who worked at the Tyler plant during the seventeen years it was in operation. And what is even more depressing is that the Paterson and Tyler tragedies are being repeated over and over, from one end of this country to the other. Last summer, as Joe and I were unearthing the environmental data on Tyler, we came across some mortality data on men who had worked in asbestos-textile plants throughout the United States. Like the Tyler data, this information had been accumulating willy-nilly in the division for years, and, incredible as it may sound, no one had seen fit to do anything about it. Just from the most cursory look at those data, almost anyone would know there had been a tragedy of immense proportions in many, if not all, of those factories. Why, the men working in them were

He explained that the committee was part of a long and complicated procedure by which criteria are developed for the recommendation of occupational-health standards. "The primary source of medical evidence and information about asbestos was provided in the NIOSH asbestos-criteria document, which I helped to write," Dr. Wagoner said. "This document included a critical evaluation of all known research on asbestos disease and a recommended standard based on this evaluation, and it was sent to Secretary James Hodgson on February 1st. The document recommends that airborne asbestos dust be controlled so that no worker is exposed over an eight-hour working day to an average of more than two fibers greater than five microns in length per cubic centimeter of air. It proposes that the two-fiber standard become effective two years after its promulgation, in order to permit manufacturers of asbestos products to install the necessary engineering controls, and that in the meantime the temporary emergency standard of five fibers remain in effect. It urges that medical surveillance, including periodic pulmonary-function tests and X-rays, be required for all workers exposed to more than one asbestos fiber per cubic centimeter of air, and that these examinations be conducted at the employer's expense. It also recommends that warning labels be affixed to containers of raw asbestos and to finished asbestos products stating that asbestos is harmful, that it may cause delayed lung injury, including asbestosis and cancer, that its dust should not be inhaled, and that it should be used only with adequate ventilation and approved respiratory devices."

Dr. Wagoner went on to tell me that in proposing a permanent two-fiber standard for asbestos dust he and the other authors of the NIOSH document gave great weight to the fact that that standard had been recommended in 1968 by the British Occupational Hygiene Society and had been adopted by Her Majesty's Inspectorate of Factories the same year. "However, we took care to point out that the British standard was designed only to reduce the early signs of asbestosis, and not to prevent asbestos-induced cancer, which may occur after exposure to levels of asbestos dust that are low enough to prevent lung scarring," he added.

Continuing, Dr. Wagoner said that the Advisory Committee on the Asbestos Standard had been set up by Secretary Hodgson two months before, in January, to provide additional evidence and information as to what the permanent standard should be. "The committee has five members, representing industry, labor, government, and the independent medical and scientific community," Dr. Wagoner said. "In addition to me, it includes Isaac H. Weaver, corporate director for environmental control of Raybestos-Manhattan, Inc.; Andrew Haas, the president of the International Association of Heat and Frost Insulators and Asbestos Workers; Jack Baliff, the chief engineer of the Division of Industrial Hygiene of the State of New York's Department of Labor; and Edwin Hyatt, of the University of California's Los Alamos Scientific Laboratory, who is the chairman. We held meetings in Washington for five days in February, and, by majority vote, we supported the two-fiber standard and all the recommendations of the NIOSH criteria document. In fact, in certain areas we made recommendations to the Secretary of Labor that were even stronger than those of the criteria document. For example, as I said, we recommended that before respirators could be issued to workers for any reason, each worker must have a complete physical examination to determine whether he could wear a respirator without endangering his health. We took this action to avoid the recurrence of conditions like those at Tyler, where respirators were slapped onto men who already had pulmonary problems as a result of exposure to asbestos."

That night, I had dinner with Dr. Johnson and his wife, who lived, with their two children, in an apartment in the suburbs of Cincinnati. I had been told that Dr. Johnson was fulfilling his military obligation by serving with NIOSH, and as he was driving me to my hotel later in the evening I asked him if he intended to remain there when his two-year tour of duty was over.

For a few moments, Dr. Johnson was silent; then he shook his head and said he really didn't know. "I am greatly troubled by the question of respectability in the field of occupational medicine," he told me. "There's very little peer pressure among the doctors who are in it, either in industry or in government, and now that I find myself faced with the problem of defining myself professionally for the next thirty years or so, I'm afraid of becoming frustrated and fatigued in this field, and of becoming part of the fabric of how things are done in a huge bureaucracy. You see, the way things are set up in occupational health these days, it's all too easy for a man to look at the welter of problems awaiting solution, to realize the lack of any real intention on the part of many people in government and in industry to take any significant action to remedy them, and to say to himself, 'Well, I can't do anything on my own, so I might just as well sit back and fit into the mold.' "

"But you did do something about it," I said. "You and Dr. Wagoner did something that could be the beginning of turning the whole thing around."

"Yes, we did something," Dr. Johnson replied quietly. "But will they let us keep on doing it?"

Early the next morning, March 14, 1972, I flew to Washington to attend the opening session of the Department of Labor's public hearings on the proposed permanent standard for occupational exposure to asbestos. They were held in a large conference room in the Interdepartmental Auditorium, at Twelfth Street and Constitution Avenue, and when I arrived there, shortly after nine o'clock, the place was filling up with some hundred-odd representatives of industry, labor, government, and the independent medical and scientific community. The morning was given over to scheduling and rescheduling appearances of people wishing to give testimony during the rest of the week, and this complicated business was accomplished with wit and dispatch by Arthur M. Goldberg, a bearded man, who was a hearing examiner for the Department of Labor. After Goldberg had arranged the agenda for the four days of hearings, a tall man in his early forties, with dark hair and white sideburns, got to his feet, introduced himself as Bradley Walls, and said he represented the Asbestos Information Association of North America. "We have a number of questions asking for rulings from you, Mr. Goldberg," he said. "I preface them by saying that, in light of the number of witnesses, we concur with you that cross-examination might delay the hearings beyond our endurance and possibly yours, and that if clarifying questions be required they best come from you, sir. Secondly, we would like your ruling on your position with regard to physical evidence, either living or photographic. We would prefer that it not be presented, inasmuch as we do not think it would be helpful to this hearing."

With a puzzled frown, Goldberg inquired, "May I ask what you mean?"

"Either basket cases or X-rays," Walls said, with a grin. "We feel that their introduction would turn the hearings into a circus."

"The only thing I can say now is that evidence must be submitted in duplicate," Goldberg said dryly.

Walls grinned again. "Thank you, sir," he replied. "We will accept that."

When Walls sat down, a slight man in his early thirties rose at the rear of the room and, in a voice full of emotion, introduced himself as Colin D. Neal, the administrative assistant to the president of the United Papermakers and Paperworkers Union, which represents twenty-one hundred workers at the Johns-Manville Corporation's asbestos plant in Manville, New Jersey. "Sir, the United Papermakers and Paperworkers would like to express our indignation at Mr. Walls's characterization of those who may suffer the effects of asbestos-dust disease as 'basket cases,' " he said. "Using his terminology, however, we have a 'basket case' we would like to present to you sometime today."

Goldberg looked at Neal and nodded slightly. Then he said, in a quiet voice, "We will hear all witnesses who are presented, sir," and adjourned for lunch.

On my way out, I encountered Sheldon Samuels, the director of Health, Safety, and Environmental Affairs for the AFL-CIO's Industrial Union Department, whom I had previously met and talked with on several occasions. Samuels, a stocky man in his middle forties, is ordinarily mild-mannered, but he was now flushed with anger. When I asked him to explain what had happened between Walls and Neal, he shook his head grimly. "We're holding a press conference at the Hotel Washington in a few minutes," he said. "Come on over and you'll find out."

The press conference was conducted by the IUD in conjunction with the United Papermakers and Paperworkers, and was attended by a dozen or so journalists from various newspapers and magazines and by a Metromedia television camera team. Seated from left to right behind a long table at the front of the room were Samuels; Dr. William Nicholson, assistant professor of community medicine at the Mount Sinai School of Medicine and a member of the Mount Sinai Environmental Sciences Laboratory; Dr. Maxwell Borow, a thoracic surgeon from Bound Brook, New Jersey, which is near Manville; Jacob Clayman, administrative director of the IUD; Colin Neal; Joseph Mondrone, president of Local 800 of the Papermakers' union in Manville; Robert Klinger, Local 800's vice-president and the chairman of its Health and Safety Committee; Daniel Maciborski, a member of the local; and Marshall Smith, the local's international representative.

Samuels got the press conference under way by reminding his listeners that it had long been known that the inhalation of asbestos dust could scar and destroy the lungs. "For the past thirty years, asbestos has been a proven cause of cancer of the lungs, and of the stomach and intestines of the workers who breathe it," he went on. "Usually, exposure over a long period of time is necessary to produce asbestos-related disease, but there is now evidence that even a single day of breathing large amounts of asbestos dust will harm the lungs. Contamination in the community, especially in the homes of asbestos workers, has been shown to cause cancer in women and children who have never been in an asbestos factory. Indeed, no one who has been or who is being exposed is safe from the effects of asbestos, and tens of thousands of workers and their families may already have had their lives shortened by exposure to asbestos dust."

Samuels went on to say that the development of safe methods of working with asbestos had been hampered for years by the efforts of management to hide the facts about asbestos disease, to suppress government and private studies of the subject, and to prevent state job-safety agencies from taking effective action. He then declared the temporary emergency standard of five fibers per cubic centimeter of air to be totally inadequate. "The Industrial Union Department will recommend at the hearings this week that a standard of two asbestos fibers per cubic centimeter of air go into effect within six months, and that within two years the standard be lowered to one fiber per cubic centimeter," he said. "Moreover, since constant monitoring of fiber levels in hundreds of plants is obviously impossible, we are calling for the installation of engineering controls and work practices designed to bring asbestos exposures ultimately to a zero level."

Samuels then introduced Clayman, who has been with the IUD since its formation, in 1956, and had been its administrative director since 1960. Clayman, a soft-spoken man in his middle sixties, has spent a lifetime in the labor movement, first as a steelworker, then as a member of the Ohio state legislature fighting for improved workmen's compensation laws, and, just before joining the IUD, as secretary-treasurer of the Congress of Industrial Organizations in Ohio. Speaking in measured tones, Clayman told his audience that the press conference had been called to bring to public attention what might well be the most devastating environmental disaster yet perpetrated by any industrial nation. "Today, millions of American workers, their families, and their neighbors may be exposed to toxic concentrations of asbestos," Clayman said. "God only knows how many thousands of workers have died, and how many will die or be terribly sick, because of the routine way this country has dealt with the problem of occupational exposure to asbestos for so many years. We cannot bring dead workers back to life or prevent pain long since experienced, but we can and must bring an end to this inexcusable environmental crime of huge proportions that affiicts workers and totally unaware victims in the plant community."

Dr. Borow was then introduced, and he described the cases of malignant mesothelioma that he and his associates at the Somerset Hospital, in Somerville, New Jersey, had begun to find in 1964, and said that he had witnessed a sharp rise in the incidence of the disease since then. He quoted from a letter he had written on October 12, 1967, to Marshall Smith, then president of the Papermakers' Local 800. The letter stated that Dr. Borow and his associates were planning an exhibit on the rising incidence of mesothelioma in the Manville area, which they had hoped to display in 1968 at four major medical conventions throughout the country and at various hospitals in New Jersey, but that, though they had applied to forty different sources for funding, they had been unable to obtain money for this purpose. "We were told frankly that local industry would not support this project for fear of upsetting the Johns-Manville Corporation," the letter continued. "Johns-Manville themselves, after six weeks of deliberation, refused support, as they were not ready to acknowledge the association between asbestosis and mesothelioma." Dr. Borow's letter to Smith concluded by asking the union to provide the three thousand dollars that would be necessary to assemble and transport the exhibit, and after he had finished reading it Dr. Borow said that the union had supplied the money and the exhibit had been widely displayed.

Dr. Borow then introduced Maciborski, a patient in whom he had discovered an abdominal mesothelioma a few months earlier. Maciborski, a gaunt man in his middle fifties, told the audience with calm and dignity that he had contracted mesothelioma while working for Johns-Manville, and that he hoped his personal misfortune would encourage government officials to act promptly so that it would not be shared by other workers.

The hearings had begun by the time I had had some lunch and returned to the conference room. As I took a seat, I saw that Maciborski and Dr. Borow had been giving testimony at a witness table at the front of the room—to the right of Goldberg, the hearing examiner, and directly opposite a cross-examination panel consisting of Nicholas DeGregorio, an attorney with the Department of Labor's Office of the Solicitor, and Gerard F. Scannell, acting director of the Occupational Safety and Health Administration's Office of Standards. Toward the end of his remarks, Dr. Borow said that he had now encountered fifty-two cases of mesothelioma in the Manville area, and that all the victims of the disease had worked for Johns-Manville with the exception of two, who had simply lived in the community.

Dr. Borow and Maciborski were followed at the witness table by Dr. Nicholson, who began his testimony by stating that the health experience of American asbestos workers could be described only as a national tragedy. Referring to the mortality study Dr. Selikoff and Dr. Hammond had made of insulation workers in the Newark-New York area, Dr. Nicholson reminded his listeners that two in ten of those men had died of lung cancer, one in ten of gastrointestinal cancer, nearly one in ten of mesothelioma, one in ten of other cancers, and almost one in ten of asbestosis. "Past standards are not an appropriate reference in setting a new permanent standard for occupational exposure to asbestos, simply because all past standards were conceived only for the purpose of preventing asbestosis," Dr. Nicholson continued. "But asbestosis is obviously not the major problem among asbestos workers. Cancer is the major problem. Cancer accounts for seventy-five per cent of the excess deaths among the asbestos-insulation workers studied by Dr. Selikoff and Dr. Hammond, and this asbestos-cancer hazard is not appropriately covered by the proposed asbestos standard." Dr. Nicholson went on to say that no knowledge now existed of a safe working level of exposure to asbestos which would prevent the occurrence of cancer, and he urged that asbestos not be used in the workplace except with approved techniques and methods designed to remove asbestos dust from the working environment. "There is evidence that a standard of two fibers per cubic centimeter of air will be inadequate for the prevention of asbestos disease," he said. "The recently measured long-term exposure of the asbestos-insulation workers, whose disastrous disease experience has been documented by Dr. Selikoff and Dr. Hammond, was approximately three fibers per cubic centimeter, even prior to the implementation of improved control measures."

Another of the afternoon's witnesses was Dr. Sidney Wolfe, who is the director of Ralph Nader's Health Research Group and a former medical researcher on the staff of the National Institutes of Health. Dr. Wolfe testified that "if workers were guinea pigs and asbestos were a food additive, the Delaney Clause of the Food and Drug Act [which prevents the introduction into the marketplace of any substance known to cause cancer in test animals] would have mandated the elimination of this carcinogenic dust from the environment long ago. However, in 1972, twelve years after the publication of data showing the relationship between asbestos exposure and mesothelioma in humans, and at a time when there are now hundreds of cases of this cancer in workers exposed to asbestos, the slaughter continues. Under these circumstances, regulations which do not ultimately reduce the fiber count to zero fail to comply with the Occupational Safety and Health Act of 1970, which clearly states that 'no employee will suffer diminished health, functional capacity, or life expectancy as a result of his work experience.'"

Dr. Wolfe was succeeded at the witness table by Anthony Mazzocchi, who was accompanied by his assistant, Steven Wodka, and who stated the position of the Oil, Chemical, and Atomic Workers International Union in blunt language. "The proposed Labor Department standard for exposure to asbestos dust is a very sad document," he said. "It serves to confirm what many members of our international union already fear—that the [Occupational Safety and Health] Administration is frivolous with the health and rights of working people." Mazzocchi went on to say that there were far more people exposed to asbestos in the workplace than one was usually led to believe. "The often quoted Labor Department figure of two hundred thousand workers isn't conservative, it's ridiculous," he declared. "In our international union, which represents one hundred and eighty thousand workers in the oil, chemical, and atomic-energy industries alone, almost every shop and plant uses asbestos in one form or another. For example, in a major oil refinery on the East Coast—Mobil Oil in Paulsboro, New Jersey—asbestos has captured our concern as the single most serious industrial-health hazard in that facility. We had nineteen workers who handle asbestos-insulation materials in that refinery examined by Dr. Irving Selikoff, of the Mount Sinai School of Medicine. Dr. Selikoff's tests revealed a very serious occupational-health problem resulting from their exposure to asbestos. Now our concern is that two to three hundred other workers—pipefitters, boilermakers, welders, bricklayers, and others who work in and around this insulation—may also have been overexposed. Asbestos turns up in the most unexpected situations. Recently, I was touring a plant in northern New Jersey where Prestone antifreeze is made. At one point in the tour, I caught a completely unprotected worker dumping asbestos into a vat of antifreeze. He told me that asbestos is what gives Prestone its anti-leak quality. If that was an unexpected situation, then what has been our experience in a primary asbestos plant—for example, one that manufactures asbestos-insulation products? Up until recently, the OCAW represented workers at the Pittsburgh Corning Corporation's asbestos plant in Tyler, Texas. This plant was the sister to the Union Asbestos & Rubber Company's factory in Paterson, New Jersey, where Dr. Selikoff conducted his now famous mortality study of amosite-asbestos workers. At the Paterson plant, Dr. Selikoff found that total deaths were more than twice the number anticipated, and now at the Tyler plant the National Institute for Occupational Safety and Health has already found that seven out of eighteen workers with ten or more years of employment meet at least three of four criteria for asbestosis. Worse yet, HEW studies of the plant dating back to 1967 have found grossly excessive levels of asbestos dust throughout the plant. While this particular factory employed only sixty or so people at its peak, the turnover was such that nearly nine hundred men had worked there for varying periods of time from 1954 to 1972. The story of Tyler is sadly filled with episodes of corporate indifference and governmental secrecy."

Mazzocchi went on to say that, because even very small quantities of asbestos were known to cause cancer, the union was recommending that all exposure to asbestos ultimately be reduced to zero by the enforcement of strict equipment-performance standards. "All manufacturing, maintenance, and other industrial and construction processes using asbestos must be re-engineered so that they perform at zero exposure," he declared. "We propose that industry be put on notice, as soon as possible, that within six months of the effective date of this standard, no worker shall be exposed to more than two fibers per cubic centimeter of air; that within two years this level shall be reduced to one fiber; and that within three years of June of 1972 zero exposure shall be the law. As for respirators, they should be authorized only when the employer has a definite abatement plan to reduce the exposure to asbestos through engineering means. The other situation in which respirators would be allowed is where there is no feasible technology for controlling asbestos dust." Mazzocchi added that the Occupational Safety and Health Administration's proposed standard on medical examinations of asbestos workers would truly allow the fox to guard the chickens. "The medical community, like many other professional groups in this country, has physicians that industry can rely on to deny valid occupational-disease claims of workers," he said. "Therefore, we recommend that workers be allowed to have annual physical examinations performed on them by doctors of their own choice, but at the employer's expense. Furthermore, the records of these examinations should not be sent to the employer but to a central record-keeping facility at NIOSH, where such records could be kept intact and confidential. NIOSH would then send each employer an annual statistical summary on the examinations of all his employees. It has been our sad experience, in case after case, that as soon as management finds out how badly it has injured the health of a worker, management does its best to get rid of him. Thus these records need to be kept intact for at least forty years." Mazzocchi concluded by declaring that a deficient standard for protection from the hazards of asbestos would legislate sickness and an early death for thousands of people. "Faced with this prospect, I would seek no new rule at all, rather than be held responsible for the cases of asbestos disease that will surface thirty years from now," he said.

One of the final witnesses of the afternoon was Alex Kuzmuk, a governor of the Asbestos Textile Institute—which in 1964 had sent a letter to the New York Academy of Sciences urging caution in the public discussion of medical research into asbestos disease in order "to avoid providing the basis for possibly damaging and misleading news stories." Kuzmuk now testified that the Asbestos Textile Institute was opposed to the NIOSH criteria document and to the recommendations of the Secretary of Labor's Advisory Committee on the Asbestos Standard. "We find that even the five-fiber standard is not feasible for us," he said. "Indeed, it will price American-made asbestos-textile products right out of the world and domestic markets, with the result that imports from nations where workers are under no such protection will flood the country. We feel that the proposed standard is based upon incomplete studies and that new evaluations are needed. Pending more comprehensive studies, we respectfully urge the Secretary of Labor to reconsider the establishment of asbestos standards, to reinstate the threshold limit value for asbestos dust at twelve fibers per cubic centimeter, and to provide for representation of the Asbestos Textile Institute on future advisory and study committees."

When Arthur Goldberg recessed the first day's session, I flew back to New York, where business kept me during the second day of the hearings. The day after that—Thursday, March 16th—I took an early plane to Washington to be present for what Goldberg had referred to previously in the proceedings as the Johns-Manville "scenario." The conference room of the Interdepartmental Auditorium was almost full when I arrived, just before 9 a.m., and the hearings got under way promptly, with John B. Jobe, Johns-Manville's executive vice-president for operations, sitting down at the witness table and stating that the asbestos industry had first supported research on asbestos disease during the nineteen-twenties, at the Saranac Laboratory of the Trudeau Foundation (in Saranac Lake, New York) and was at present supporting such research at more than half a dozen medical schools in the United States and Canada. He went on to say that although the asbestos industry recognized its responsibility to support the intent of the Occupational Safety and Health Act, there was no credible evidence demonstrating the necessity for a standard lower than five fibers per cubic centimeter of air.

Jobe was followed by Dr. George W. Wright, a longtime paid medical consultant for Johns-Manville, who was also director of medical research of the Department of Medicine of St. Luke's Hospital in Cleveland. Dr. Wright began his testimony by saying that he had been conducting research on asbestosis since 1939, first as a member of the Saranac Laboratory of the Trudeau Foundation and then, since 1953, at St. Luke's Hospital. After reviewing the various standards for occupational exposure to asbestos that had been in effect over the years, Dr. Wright told the hearings that no evidence had been found to indicate that the present asbestos standard should be changed. "Moreover, since I believe that the five-fiber standard will certainly prevent asbestosis, I am in complete disagreement with the NIOSH criteria document with respect to its expressed opinion that the data relating asbestos exposure to biological reaction are inadequate to establish a meaningful standard at this time," he said. "While the evidence may not be as far-reaching as we would like, it is scientifically valid, and adequate to support as a first approximation the opinion that the present standard of five fibers per cubic centimeter should not be lowered, but left as it is."

According to Dr. Wright, a recent study conducted by Dr. John Corbett McDonald, of the Department of Epidemiology and Health of McGill University, in Montreal, furnished strong support for not lowering the asbestos standard below five fibers per cubic centimeter of air, and proof that mesothelioma was virtually absent in people who were exposed only to chrysotile asbestos—a type of the mineral that accounts for ninety-five per cent of the world's production, and the type that Johns-Manville mines, uses, and sells almost exclusively. "Mesothelioma appears to be predominantly linked with exposure to crocidolite or amosite," Dr. Wright declared. "Therefore, both of these types of asbestos should be controlled more stringently than is chrysotile." Dr. Wright then criticized certain aspects of Dr. Selikoff's and Dr. Hammond's mortality studies of the asbestos-insulation workers; the studies did not include adequate control populations, he said, and the incidence of mesothelioma among these workers was caused not by their exposure to chrysotile but by their dual exposure to chrysotile and amosite. He ended by reiterating his support of the five-fiber standard, because, as he put it, "this is a correct standard and constitutes a level of exposure that will protect against the development of asbestosis and bronchogenic cancer."

Thus far in the hearings, there had been very little cross-examination, but when Dr. Wright concluded his remarks a number of people made it known that they had questions to ask and points to make concerning his testimony. Among them was Nicholas DeGregorio, of the Department of Labor, who pointed out with some asperity that he had never heard the validity of Dr. Selikoff's and Dr. Hammond's study of the asbestos-insulation workers questioned by any of the leading epidemiologists in the field.

After a short recess, the Johns-Manville testimony continued with the appearance at the witness table of Dr. Thomas H. Davison, who introduced himself as the medical director of the corporation. Dr. Davison's testimony was very brief, and was chiefly concerned with his objections to the proposed frequency of medical examinations for asbestos workers. When he completed his remarks, he was succeeded at the witness table by Edmund M. Fenner, the corporation's director of environmental control. Fenner testified that Johns-Manville had worked diligently to lower dust levels in all its plants. He also criticized the two-fiber standard proposed in the document, on the ground that adequate monitoring and dust-sampling equipment was not available to measure such a level.

Then Dr. Fred L. Pundsack, Johns-Manville's vice-president for research and development, came to the witness table. "Perhaps nowhere else in the asbestos standards being considered today is the opportunity to bring about bad changes so clearly evident as it is in some of the proposed label requirements," Dr. Pundsack said. "If these label requirements are adopted in their proposed form, they will in our opinion destroy large amounts of the industry and eliminate thousands of jobs." Dr. Pundsack went on to declare that warning labels need only indicate that precautionary steps should be taken when handling asbestos, and that labels need not contain terrifying language, such as the word "cancer." He pointed out that asbestos is not an acutely toxic chemical or drug that reacts within minutes or hours, nor is it an explosive, nor can it be absorbed through the skin. "Therefore, the application of frightening labels to asbestos is inappropriate," he said. "Instead, we recommend that a caution or warning label with the following type of text be used on bags or containers of asbestos fiber: 'Caution—This bag contains chrysotile-asbestos fiber. Inhalation of asbestos in excessive quantities over long periods of time may be harmful. If proper dust control cannot be provided, respirators approved by the United States Bureau of Mines for protection against pneumoconiosis-producing dusts should be worn.' "

When Dr. Pundsack finished his remarks, there was an hour's recess for lunch. The first afternoon witness was Henry B. Moreno, senior vice-president for the industrial and international divisions of Johns-Manville, who said that the company's dust-control programs had already cost twenty million dollars. "For us to achieve a standard of two fibers per cubic centimeter would require capital expenditures of twelve million dollars, and additional operating costs of five million dollars per year," Moreno declared. "It would simply not be economically feasible to operate at this level in five of our plants, which, if closed down, would put sixteen hundred employees out of work. This and similar closings across the country would have a substantial effect upon the nation's economy, and would result in higher costs reflected all across the board. In addition, Japan, Taiwan, India, other Asian countries, and nations in South America would come on strong and flood the American market with asbestos products. For these reasons, we believe that it would be nothing less than complete social irresponsibility to adopt a two-fiber standard for occupational exposure to asbestos without stronger medical evidence than that which presently exists."

When questioned by Dr. Nicholson, Moreno, like Dr. Wright before him, sought to place chrysotile asbestos above suspicion as a cause of mesothelioma, and, like Dr. Wright, he implicated amosite. Moreno declared that from 1930 until 1960 all high-temperature-insulation materials contained amosite, that since 1960 there had been a trend away from amosite, and that for the past five years almost no amosite had been used.

Knowing that Johns-Manville had long been attempting to absolve chrysotile by blaming crocidolite and amosite asbestos for the occurrence of mesothelioma, and that most members of the independent medical and scientific community considered such efforts to be self-serving, I was not surprised to hear Dr.

Nicholson strongly question Moreno about his statement that amosite asbestos had been a major constituent of insulation materials between 1930 and 1960. Later, I learned that Dr. Nicholson reinforced this refutation by sending an addendum to Goldberg on March 24th for inclusion in the record of the hearings. Dr. Nicholson's accompanying letter referred Goldberg to two tables of information he had included in his addendum. The first table, which listed the quantity of asbestos used in the manufacture of insulation materials in the United States between 1920 and 1965, had been furnished by Dr. Pundsack himself to Dr. Selikoff for presentation at the Fourth International Pneumoconiosis Conference of the International Labour Organisation, held in Bucharest, on September 29, 1971. The second table, compiled from the *United States Minerals Yearbooks*, listed imports of amosite asbestos into the United States during those same years. Since a comparison of the two tables showed that only a few hundred tons of amosite was imported each year between 1920 and 1940, and that this amount was only a small fraction of the total amount of asbestos used in the manufacture of insulation materials during that period, Dr. Nicholson pointed out, "clearly, amosite could have been only a minor constituent of insulation material until World War II," and even through 1950 "it could only represent a small fraction of the asbestos used in nonmarine commercial and industrial insulation, if one considers the extensive use in shipbuilding." Dr. Nicholson concluded his letter by calling Goldberg's attention to a table showing that the disease experience (including mesothelioma) of shipyard insulation workers was not significantly different from the disease experience of nonshipyard insulation workers. "It is not possible to assign an important role to amosite in the insulation workers' experience," he wrote.

After Dr. Nicholson's cross-examination of Moreno, the seat at the witness table was taken by Dr. McDonald, who stated at the outset that he was a professor of epidemiology and the chairman of the Department of Epidemiology and Health of McGill University, and that he had specialized in epidemiology for twenty-four years. "I would now like to add one or two points not in my written submission, in order to clarify my position here," Dr. McDonald continued. "The first point is that I am a full-time employee at McGill University, and an independent research worker. I do not work, nor am I associated, with any asbestos producer or manufacturer. The research I shall be describing is supported by grants, not to me but to McGill University, from a number of sources—the Institute of Occupational and Environmental Health, the Canadian government, the British Medical Research Council, and the United States Public Health Service. I am not here to support the testimony or position of Johns-Manville or any other body affected by the proposed regulations."

Dr. McDonald went on to quote at length from a report entitled "The Health of Chrysotile Asbestos Mine and Mill Workers of Quebec," which he and some colleagues were preparing for publication in the near future. Dr. McDonald said that he and his associates had begun an epidemiological study of miners and millers in 1966, using records of the Quebec asbestos-mining companies to identify all persons known to have worked in the industry since its inception in 1878. He explained that the mortality aspect of the study was limited to those men who had worked for a month or more, and who were born between 1891 and 1920, adding that he and his colleagues had already published an initial analysis of the mortality experience of these workers. Dr. McDonald then said that about eighty-seven per cent of the 11,572 persons included in the mortality study had been traced by the end of December 1969, and that 3270 of them had died. "Cancer of the lung showed a rising death rate with increasing dust exposure, particularly in the two highest dust-exposure groups," he continued. "Of one hundred and thirty-four deaths from respiratory cancer, there were five from pleural mesothelioma. These cases, however, showed no clear relationship with dust exposure."

Later in his presentation, Dr. McDonald assessed the results of his mortality study by declaring that the number of excess deaths related to asbestos exposure among the workers he had investigated probably constituted no more than two per cent of the total of 3270 deaths; that most of these deaths were caused by lung cancer and pneumoconiosis (by which he presumably meant asbestosis); and that almost all of these excess asbestos-related deaths occurred among workers employed in the highest dust-exposure categories. After pointing out that the death rates from cancer and mesothelioma among the chrysotile-asbestos miners and millers he had studied were very low compared with the death rates from those diseases found among the insulation workers studied by Dr. Selikoff and Dr. Hammond, Dr. McDonald concluded that only high levels of exposure to chrysotile asbestos during mining and milling operations had an appreciable effect on mortality. Dr. McDonald ended his presentation by further concluding, from the findings of his study, that a reasonable standard for chrysotile mines and mills would be somewhere between five and nine fibers per cubic centimeter.

When Dr. McDonald finished his testimony, he was questioned at some length by Dr. Nicholson and by DeGregorio. Dr. Nicholson's questioning elicited a statement from Dr. McDonald that in a previously published report on mortality among the Quebec asbestos miners and millers, he had concluded that among those workers in his study who were exposed to the highest level of chrysotile dust the incidence of lung cancer was five times that of the workers exposed to the lowest level. He also obtained an admission from Dr. McDonald that his recommendation of a standard of between five and nine fibers was based upon a total of only thirty-two fiber counts made in mines and mills of Quebec in the summer of 1971. DeGregorio, too, asked Dr. McDonald a series of pointed questions about the scientific validity of his study. He expressed open skepticism of Dr. McDonald's ability to substantiate the accuracy of chrysotile-exposure levels that workers were exposed to during the nineteen-fifties and the nineteen-sixties. He also obtained an admission from him that not all the effects of whatever exposures there may have been were observed directly by Dr. McDonald and his associates—through, for example, the examination of autopsy material—but that they had been observed by other people and recorded by them in reports and death certificates, which he and his associates had then included in their study as valid.

I was not surprised to hear Dr. McDonald questioned in this manner, for several members of the independent medical and scientific community had previously expressed grave reservations to me about the accuracy of the conclusions he and his colleagues had drawn in a report of their study which had appeared in June of 1971, in Volume XXII of the *Archives of Environmental Health*, under the title "Mortality in the Chrysotile Asbestos Mines and Mills of Quebec." Some people had pointed out that many, if not most, of the workers studied by Dr. McDonald could have had little or no exposure to airborne asbestos fibers, because they had worked in open-air pits, extracting asbestos in wet-rock form. Others deplored the fact that Dr. McDonald and his associates had conducted very little pathological review, such as the examination of autopsy material and lung-tissue slides, in arriving at their conclusions. Still others pointed out that ninety per cent of the lung cancers and mesotheliomas found in insulation workers occurred twenty years or more after the onset of exposure to asbestos—as, for example, in the cases of men who began working with asbestos at the age of twenty, and who died of cancer at fifty—and that by omitting persons born before 1891 Dr. McDonald and his associates had excluded from their calculations precisely the people who might be expected to show the effects of asbestos inhalation. (It was as if in studying the total occurrence of gray hair one refused to look at anyone born more than forty or fifty years ago.) In addition, a number of people pointed out that by including only deaths that occurred twenty years or less after the onset of exposure, Dr. McDonald had perforce diluted the major disease effect of asbestos in his study. Perhaps the most telling criticism of Dr. McDonald's study, however, was made in a letter sent to Dr. Selikoff on January 7, 1972, by Herbert Seidman, who is chief of statistical analysis in the Department of Epidemiology and Statistics of the American Cancer Society. Seidman's critique was included in the addendum for the hearing record that was submitted by Dr. Nicholson. It described some of Dr. McDonald's methods of computing death rates as "ill-advised." It pointed out the lack of consideration that Dr. McDonald and his associates had given to the importance of the long latency period in the development of asbestos tumors, and it described the methodology used in the study to assess separately the importance of cumulative dust exposure and duration of exposure in relation to lung cancer as "inappropriate," because of the "paucity of basic data." In conclusion, Seidman wrote, "I think that the data have been collected fairly well but analyzed quite poorly."

As a layman, I had little way of judging the scientific validity of Dr. McDonald's work except through the observations of those members of the independent medical community who had communicated their opinions of it to me. However, I had brought with me to the hearings a copy of Volume XXII of the *Archives of Environmental Health*, containing Dr. McDonald's article on mortality among the chrysotile-asbestos miners and millers of Quebec, which had been sent to me some months earlier by William P. Raines, a vice-president and director of public affairs for Johns-Manville. Since Dr. McDonald had referred to this mortality study during the course of his testimony, and since anyone attending the public hearings had the right to cross-examine witnesses, including members of the press, I decided to ask him some questions about it. After receiving permission from Goldberg to address Dr. McDonald, I reminded him that in his opening remarks he had declared that all his research had been performed independently.

"That is correct," Dr. McDonald replied. "All things are relative."

I then reminded Dr. McDonald that John Jobe had testified at the morning session that his company was supporting research on asbestos disease, and asked him if that was research other than what he had performed.

"I guess what Mr. Jobe is referring to is the fact that Johns-Manville, together with other mining companies, helps support the Institute of Occupational and Environmental Health, which is a granting body that receives research applications, and which therefore indirectly supports our research," Dr. McDonald replied. "Now, it is a very indirect relationship."

I then pointed out to Dr. McDonald that at the end of his article in the *Archives of Environmental Health*, a credit was listed in small type: "This work was undertaken with the assistance of a grant from the Institute of Occupational and Environmental Health of the Quebec Asbestos Mining Association."

"That is correct," Dr. McDonald said.

With that, I took my seat. Dr. McDonald had just indirectly admitted that Johns-Manville, together with other asbestos-mining companies, supported the Institute of Occupational and Environmental Health, and that the institute, in turn, had helped support his study. Moreover, the credits at the end of his article, which listed no financial support other than that supplied by the institute, had given the full and correct title of this organization—the Institute of Occupational and Environmental Health of the Quebec Asbestos Mining Association. It seemed unnecessary to point out to the representatives of industry, labor, government, and the independent medical and scientific community who were gathered in the conference room something that many of them already knew—that Johns-Manville is, and for the past quarter of a century has been, the dominant member of the Quebec Asbestos Mining Association.

When the hearings were adjourned that afternoon, Ivan Sabourin, former attorney for the Quebec Asbestos Mining Association, came up to me and introduced himself. We talked briefly, and then I took a plane back to New York. I had never met Sabourin before, but I remembered reading something about him in connection with McGill University in a copy of the minutes of a 1965 meeting of the Asbestos Textile Institute. The following day, I took the minutes from my files and read them again. They informed me that a meeting was held on June 4, 1965, at the Motel Le Provence, in Thetford Mines, Canada, and they quoted Sabourin as saying that a recent article associating asbestos and cancer in the *Journal of the American Medical Association* was not convincing, and expressing regret over the adverse publicity that resulted from such articles. Sabourin then told the meeting that the Quebec Asbestos Mining Association wished to study respiratory diseases related to chrysotile asbestos, and that it was seeking "alliance with some university, such as McGill, for example, so that authoritative background for publicity can be had."

According to the minutes, the next speaker at the meeting was Dr. Lewis J. Cralley, of the United States Public Health Service, "who for the past several years has been supervising the extensive environmental study of asbestos employees in textile plants in the U.S.A." Dr. Cralley told the meeting that "the study was going well," that the Public Health Service was now extending its work into other asbestos industries, and that "the results to date certainly justify the program and its further expansion." Dr. Cralley did not elaborate on what these results had been, nor, for that matter, did he ever see fit to officially warn any segment of the asbestos industry, least of all the workers, that the data he was collecting showed that men employed in asbestos factories across the land were being exposed to grossly excessive levels of asbestos dust, and that excess mortality from asbestos disease among workers in asbestos-textile factories had reached tragic proportions. (Indeed, six years passed before Dr. Johnson and Dr. Wagoner unearthed the data buried in Dr. Cralley's files and undertook to do something to rectify the appalling situation they discovered.) In this connection, I found it interesting to note that out of the seventy-odd people listed in the minutes as attending the 1965 meeting of the Asbestos Textile Institute, Dr. Cralley was the only invitee from any government, and the only one who did not represent an asbestos company or a related organization.

At the same time, I also reread a paper sent to me some months before by Johns-Manville, which gave a history of the company's health-research programs. Referring to Dr. McDonald's study of the Quebec asbestos miners and millers, the paper had this to say:

> This study is being funded by the Institute of Occupational and Environmental Health, the scientific research arm of the Quebec Asbestos Mining Association (QAMA). As mentioned before, Johns-Manville is a principal member of the QAMA. The Institute of Occupational and Environmental Health plays a vital role in the Johns-Manville health research effort. Besides allocating QAMA funds for research projects, the seven-man scientific advisory committee of the Institute also reviews requests J-M receives from scientists and scientific organizations for money to conduct research in the asbestos/health field.

The paper then listed the chairman of the Institute's seven-man scientific advisory committee as Dr. George W. Wright, Director of Medical Research, St. Luke's Hospital, Cleveland, Ohio.

I did not return to Washington for the final day of the hearings, but during the following week, thanks to Gershon Fishbein, editor of the *Occupational Health & Safety Letter*, and as a result of reading the *Occupational Safety & Health Reporter*, a newsletter published by the Bureau of National Affairs, Inc., I was able to keep abreast of most of the testimony that had been delivered during the two days of hearings I missed. By and large, this testimony ran true to form, in that it reflected the beliefs and self-interest of those who delivered it. Representatives of the asbestos industry, on the one hand, stated that an asbestos standard of two fibers per cubic centimeter of air either could not be achieved technically or would be prohibitive in cost, and that it would surely result in the shutting down of many asbestos-manufacturing plants, with an attendant loss of jobs and an influx of foreign asbestos products into the United States. Representatives of labor unions, on the other hand, urged that the safety and health of workers be placed ahead of any economic considerations, that the two-fiber standard be adopted, and that efforts be made to reduce occupational exposure to asbestos to zero. In a way, much of this testimony tended to be misleading, for the hearings on the asbestos standard had become far more than just a disagreement between industry and labor over whether the standard should be five or two fibers. The introduction—by Dr. Selikoff and his associates at the Mount Sinai Environmental Sciences Laboratory, by the authors of the NIOSH criteria document, and by the Secretary of Labor's Advisory Committee on the Asbestos Standard—of proposals for performance standards that would *a priori* reduce dust levels in the manufacturing and installation of asbestos products by requiring the use of proper equipment, efficient exhaust and ventilation systems, and safe work practices was of crucial importance, for the carrying out of performance standards would obviously put the horse before the cart, where it belonged. In short, effective performance standards would be bound to lessen the importance of and reliance upon the laborious and time-consuming process of taking air samples and counting asbestos fibers beneath a microscope in order to determine whether the asbestos standard was being complied with. Thus, performance standards would go a long way toward obviating the kind of cooperation between industry and government that in factories such as Pittsburgh Corning's Tyler plant had for so many years reduced the taking of air samples and the counting of asbestos particles and fibers to a farce of tragic proportions and fatal consequences.

Two pieces of testimony delivered at the sessions I had missed were of particular interest to me in this respect, so a few days after the hearings were concluded I obtained full texts from the men who had presented them. The first was given on the second day by Duncan A. Holaday, research associate professor at the

Mount Sinai Environmental Sciences Laboratory and formerly a senior industrial-hygiene engineer with the United States Public Health Service, where he had been instrumental in developing standards for protecting uranium miners against radiation exposure. (For this work, he had been given the Distinguished Service Award of the Health Physics Society.) Holaday addressed himself at the hearings to the problem of how best to control asbestos dust:

> The use of procedural standards, by which I mean regulations requiring the use of specified methods of treating and packing material, and work rules that reduce dust production and dispersion, is the best means of preventing overexposures to harmful substances. It is based upon the knowledge that certain operations and processes will release contaminants in the work area unless they are controlled. It is also known from experience that certain control measures will markedly reduce or eliminate these emissions. Therefore, the prudent course is to require that control procedures be instituted without waiting for information obtained by air samples and dust counts to demonstrate that contamination has, in fact, occurred.

The second piece of testimony that I found of special interest was delivered on the final day of the hearings by Sheldon Samuels. He began by saying that there were certain advantages in appearing at the end of the prolonged hearings. "As you know, Mr. Goldberg, I did not plan it that way, but it has provided me with an important overview, which I intend to exploit," he declared. "The basic issue before us was made crystal clear at your prehearing conference, when Mr. Walls, of the Asbestos Information Association, attempted to prevent Daniel Maciborski from being heard, and referred to him in a disgustingly unmentionable manner. Daniel Maciborski did not ask to be heard at these hearings for dramatic effect. He was trying to tell you that more than the company's admittedly advanced environmental-control and medical-surveillance programs were needed to reduce the risk to other workers. The issue before us is whether human life can be traded off in the marketplace, and whether workers must really face death on the job."

Samuels continued his testimony by urging the adoption of performance standards that would require equipment and work practices designed for zero emission of asbestos. "For a six-month transitional period the Industrial Union Department recommends a two-fiber level," he said. "Within two years, this level should be lowered to one fiber per cubic centimeter of air, and, ultimately, there should be a zero exposure to asbestos dust." Samuels also urged the adoption and strengthening of the NIOSH recommendations for labeling asbestos, for monitoring airborne asbestos dust, for conducting periodic medical examinations of asbestos workers, and for guaranteeing that the records of such examinations be the property of the employee, and not the employer. "Most important of all, any employee who lacks confidence in the judgment of a physician who is directly responsible to the employer should have the right to choose another source of medical service," Samuels declared, adding that Maciborski had passed a medical examination provided by a Johns-Manville physician only a few weeks before his own physician had diagnosed him as suffering from terminal mesothelioma.

Most of the members of the independent medical and scientific community with whom I spoke seemed pleased by what had taken place at the hearings, and thought it likely that a two-fiber level would be adopted by the Occupational Safety and Health Administration as a permanent standard for occupational exposure to asbestos. Their optimism was based largely upon the reasoning that except for the testimony of Dr. Wright and Dr. McDonald—neither of whom could be considered completely independent medical researchers—the asbestos industry had set forth no real data to refute the conclusions and recommendations of the NIOSH criteria document and the Secretary of Labor's Advisory Committee on the Asbestos Standard. Mazzocchi, Samuels, and other union people, however, expressed a skepticism concerning the Department of Labor's motives and intent which was based upon long and bitter experience. In any event, once the hearings were concluded, nobody involved in the matter could do much but wait until June 6th, when, having presumably weighed all the evidence, the Department of Labor was required by law to promulgate a permanent standard for asbestos.

On Monday morning, March 20th, I received a long-distance call from James Bierer, the president of Pittsburgh Corning. Bierer started out by apologizing for not getting back to me sooner concerning my request to interview Dr. Grant about the Tyler plant. Then he told me that, upon the advice of legal counsel—because of the recent hearings in Washington and on account of possible litigation inherent in the Tyler situation—Pittsburgh Corning could not authorize me to conduct an interview with Dr. Grant, or, for that matter, with anyone else in its employ.

During the first week of April, I drove out to Paterson, New Jersey, and spent a day at the offices of the Mount Sinai School of Medicine's Paterson Asbestos Control Program, where Dorothy Perron and several aides (among them Shirley S. Levine, Rayla Margoles, and Charles Nolan) have been working since 1968 to trace the nine hundred and thirty-three men who had worked between 1941 and 1945 at the Union Asbestos & Rubber Company's plant there. I learned that, pressed by its insurance company, Union Asbestos had paid its workers five cents an hour extra to wear respirators, and had threatened in editorials published in the plant newspaper to fire them if they refused. I also discovered that the workers had lodged numerous complaints about the respirators, saying that they were difficult to breathe through. Indeed, some of the men had complained that, unable to work with the respirators, they had coated their nostrils with Vaseline and drunk large quantities of milk in an attempt to protect their respiratory tracts from the irritating amounts of airborne asbestos dust that filled the plant. (Obviously, such measures were pitiful protection against the pervasive nature of asbestos fibers, for when Dr. Selikoff and Dr. Hammond conducted their study of mortality among the men who had worked in the plant, they found a gross number of excess deaths resulting from asbestosis, lung cancer, mesothelioma, and other malignant tumors. Moreover, the asbestos-disease hazard extended far beyond workers directly involved in the production of insulation materials. For example, Rudolph Wild, the engineer who had developed the product manufactured in the Paterson and Tyler plants, died of mesothelioma. He may have had ample occupational exposure to asbestos, but his daughter also died of mesothelioma, and her only known exposure to asbestos had occurred when as a child she had played with samples of asbestos products her father had brought home from work. In addition to the engineer and his daughter, Robert E. Cryor, who had been manager of the Paterson plant between 1941 and 1944, died of mesothelioma in April of 1970.) During my visit to the Paterson Asbestos Control Program, I went through nearly fifty separate reports of medical examinations conducted by the company's physician which either told of abnormal lung X-rays or contained such notations as "This man is a poor risk" and "This man should not be put into a dusty area." I also discovered that during the war all blacks hired at the Union Asbestos plant in Paterson were automatically assigned to the shipping department, where dust levels were considerably lower than on the production lines, because of a belief—widely held at the time—that the lungs of black people were somehow more susceptible to dust than the lungs of whites.

While I was in Paterson, I called Thomas Callahan, of Waldwick, New Jersey, who had been a foreman in charge of the asbestos-block department of the Paterson plant. Callahan had been sent to Tyler in October of 1954 to help set up machinery for the new factory that Union Asbestos was opening there and that was later purchased by Pittsburgh Corning. "I stayed a couple of months in Tyler, and then I was sent to the Union Asbestos plant in Bloomington, Illinois, where I worked for the next eight years," Callahan told me. "As far as I was concerned, our biggest problem was health. I always wore a respirator at Paterson, at Tyler, and up in Bloomington, and on one occasion I discharged a man who refused to wear his. A lot of men hated to wear them, you know. None of them seemed to understand the hazard."

Callahan went on to tell me that he felt that the Union Asbestos people had been concerned about the safety and health of the workers in the Paterson plant. "The company doctor X-rayed all the men continually to detect asbestosis, and, once he suspected it, he would always tell a fellow to get himself a job out-of-doors," Callahan said. "In addition, the company used to pay its workers an extra five cents an hour to wear their masks, but the men were human beings, you see, and a lot of them wouldn't conform to regulations. Believe me, the company did everything it could in those days, but there was no way it could improve the ventilation system. In any case, we were a lot more humane than other people in the business. I remember going one day in the early fifties with Edward Shuman—he was then the general manager of the plant—to see some Johns-Manville people in New York. We asked them if they knew of any way we could improve the dust situation in our factory. My God, they were brutal bastards! Why, they practically laughed in our faces! They told us that workmen's-compensation payments were the same for death as for disability. In effect, they told us to let the men *work themselves to death*! Afterward, we went to the Metropolitan Life Insurance people. Only one doctor over there knew anything about asbestosis. He told us that the only solution was to spot it early and tell the guy to run for his life. We did our best, you understand, but a lot of the men wouldn't wear their respirators, and our engineers told us it was impossible to improve the ventilation."

The next day, I dropped by the Mount Sinai Environmental Sciences Laboratory to see what progress Dr. Selikoff and Dr. Hammond had made in their investigation into the mortality experience of the Paterson workers. Dr. Selikoff told me that as of December 31, 1971, Mrs. Perron and her associates had been able to trace eight hundred and seventy-seven of the nine hundred and thirty-three men who had worked at the Paterson plant during the war years. "It was a remarkable job of detective work, and Charles Nolan in particular has been incredibly adept at tracking down men who appeared to have dropped from sight," Dr. Selikoff said. "On the basis of the standard mortality tables, Dr. Hammond has calculated that in a normal population of that size, two hundred and ninety-nine deaths were to be expected. Instead, there were four hundred and eighty-four. As with the studies we conducted of the asbestos-insulation workers, the reason for the excess deaths—one hundred and eighty-five, in this case—was not hard to come by. There should have been about fifty deaths from cancer of all sites. Instead, there were a hundred and forty-three. Only eleven of the men could have been expected to die of lung cancer, but there were actually seventy-three—a rate that is almost seven times as high as that of the general population. And though virtually none of these workers could have been expected to die of mesothelioma according to the mortality tables for the general population, there were seven deaths from the disease. Moreover, in this group of men the death rate from cancers of the stomach, colon, and esophagus were twice as high as they should have been. And though none of the men could have been expected to die of asbestosis, twenty-seven of them did."

When I asked Dr. Selikoff how he felt these statistics for the Paterson workers applied to the eight hundred and ninety-five men who had worked at the Tyler factory between 1954 and 1972, he shook his head. "I can only say that for the younger men—those who could be expected to live from twenty to fifty years after their first exposure to asbestos—the future looks awfully dismal," he replied.

Dr. Selikoff then told me that in 1971 Local 800 of the United Papermakers and Paperworkers Union had asked him and Dr. Hammond to review the medical histories of its members to help evaluate the effectiveness of Johns-Manville's dust-control measures at its Manville plant. "We have since completed this study, which, sadly, serves to corroborate our previous findings," he said, adding that he would call Dr. Nicholson in and let him describe the actual results, since he had headed the field team that developed the information.

Dr. Nicholson told me that out of a total of three thousand and seven employees at the Manville complex of factories he, Dr. Selikoff, and Dr. Hammond had decided to review the histories of the six hundred and eighty-nine production workers who were actively at work on January 1, 1959, and had by that time had at least twenty years' exposure to asbestos. "We studied the mortality experience of these men from January 1, 1959, until December 31, 1971," Dr. Nicholson said. "Unhappily, the results were at least as depressing as those for the Newark-New York asbestos-insulation workers and for the men employed in the Paterson plant. Using standard mortality tables of the National Center for Health Statistics, Dr. Hammond calculated that one hundred and thirty-four deaths were to be expected in this group of people. Instead, there were a hundred and ninety-nine."

Dr. Nicholson went on to say that the reasons for this large number of excess deaths among the Johns-Manville workers were, unfortunately, all too familiar. "Only eight deaths from lung cancer should have occurred, but there were twenty-seven," he told me. "And though no deaths from mesothelioma could normally be expected, there were fifteen. Cancers of the stomach, colon, and rectum were two and a half times what they should have been. In addition, though virtually no deaths from asbestosis would have been expected among the general population, twenty-four of the Johns-Manville employees died of this disease."

When I asked Dr. Selikoff if he thought it likely that the proposed two-fiber level would be adopted by the Department of Labor as a permanent standard for occupational exposure to asbestos, he shrugged. "I have no idea," he replied. "There has been a strange development in the past week that leads me to wonder, but before I tell you about it, I'd like Dr. Nicholson to give you his outlook on number standards in general, for I wholeheartedly concur with it."

"I tend to think of number standards in this way," Dr. Nicholson said. "A standard specified as two fibers per cubic centimeter of air or five fibers per cubic centimeter of air sounds fairly innocuous. However, it is well to remember that a worker may inhale eight cubic meters, or eight million cubic centimeters, of air in a working day. Leaving aside the fact that there are many more fibers smaller than five microns in length in any environment containing airborne asbestos dust, a five-fiber-per-cubic-centimeter standard thus becomes, in terms of a man's lungs, a forty-million-fiber-a-day standard, and by the same token the proposed two-fiber standard would allow a worker to inhale sixteen million fibers a day. This, you see, is why we testified at the hearings in favor of performance standards designed not only to control asbestos emissions but to reduce them as close as possible to zero."

When Dr. Nicholson had concluded, I asked Dr. Selikoff to tell me about the recent development that had caused him to wonder whether the Department of Labor would promulgate the proposed two-fiber standard. By way of reply, he handed me a set of documents that included a page with this heading:

ENCLOSURE A
EXPERT JUDGMENTS: ASBESTOS
MEDICAL & INDUSTRIAL HYGIENE

Beneath this was a request: *"Return as soon as possible to Arthur D. Little, Inc., 35 Acorn Park, Cambridge, Massachusetts, 02140. Retain a copy for reference during Phase II."* Beneath the request was a space for the name and affiliation of the person to whom the documents were sent, and beneath that, under the words "Exposure-Response Judgments," was a table of boxes that asked the recipient to estimate what might be the incidence of asbestosis, lung cancer, and mesothelioma in a hundred workers after forty years of exposure, on the basis of an eight-hour working day, to two, five, twelve, and thirty asbestos fibers per cubic centimeter of air. Dr. Selikoff had filled in the boxes, and these were his estimates: At two fibers per cubic centimeter, fifty-five of a hundred workers would contract asbestosis, twelve of a hundred would develop lung cancer, and four of a hundred would be afflicted with mesothelioma. At five fibers per cubic centimeter (basing his judgment on what had happened to the asbestos-insulation workers), eighty-five of a hundred would develop asbestosis, twenty of a hundred would contract lung cancer, and seven of a hundred would develop mesothelioma. Dr. Selikoff's estimates were, of course, higher for workers exposed to twelve fibers per cubic centimeter of air, and for workers exposed to thirty fibers he estimated that ninety-five of a hundred would be afflicted with asbestosis, twenty of a hundred would be afflicted with lung cancer, and five of a hundred would develop mesothelioma. The reason Dr. Selikoff estimated fewer mesotheliomas at the highest level of exposure to asbestos dust was simply that previous studies had indicated that there would be more early deaths from asbestosis at such levels, and that fewer individuals would, therefore, survive long enough to develop mesothelioma.

The Arthur D. Little Phase I questionnaire also asked for a judgment on how frequently asbestos workers should be examined, and it stated that all the estimates and judgments solicited would be synthesized and included in a Phase II questionnaire, which would be sent out later. The front page of the questionnaire, which was headed "Health & Asbestos, Phase I Judgments, Background," explained what the Arthur D. Little people had in mind:

> The formulation of public policy for coping with an occupational hazard such as asbestos will necessarily rely upon judgment until a great deal more research evidence is available than now exists. In particular, judgment concerning the relationship between exposure and response will be implicit in health standards for asbestos established by the Occupational Safety and Health Administration in the near future. But judgments, possibly different ones, on the same issue will be implicit in the response of labor and of

> industry to the proposed standards. As long as judgments on the response to exposure relationship are implicit rather than explicit and as long as groups affected by the standard lack needed data to buttress their judgments, protracted conflicts are inevitable and difficult to resolve. Moreover, the absence of a clearly defined and credible set of judgments makes it difficult, if not impossible, to identify the various costs and benefits associated with policies for reducing the hazard. This is so because the benefits of candidate standards depend upon projection of lives saved or illnesses eliminated at various exposure levels. So crucial a matter should not depend upon implicit judgment or even the explicit view of a single expert. We are led to a search for a consensus that will make explicit and credible the necessary judgments on the exposure-response relationship for asbestos. Such a consensus is sought through the participation of 12 to 15 qualified experts whose judgments will be obtained, combined, and refined in a systematic way—a variant of the Delphi process that has been used extensively to apply expertise to important issues not yet open to analysis.

When I told Dr. Selikoff that I had never heard of the Delphi process, and asked him what it meant, he shook his head and smiled. "I've never heard of it, either," he said. "But I'm pretty sure I know what it means. It means guesswork. And what's the point of guessing about the biological effects of asbestos when mortality studies of asbestos workers have already shown exactly what the effect has been?"

Dr. Selikoff now handed me a letter he had written on April 3rd to Mrs. Sonja T. Strong, of Arthur D. Little, Inc., concerning the Phase I questionnaire. Regarding the effectiveness of dust-counting as a method of insuring safe working conditions, Dr. Selikoff wrote:

> As matters now stand, meager use of performance standards seems to be intended. In this case, nothing in our experience indicates that the threshold limit values listed in your questionnaire will provide any effective safeguard against the occurrence of disease.
>
> An obvious rejoinder might be: "Yes, but what if they

> were enforced? How much disease might then occur?" Following you into this never-never land, in which one-tenth of the workmen wear personal samplers on their coveralls and the rest of us are at the phase microscopes in the laboratory, the results would still not be very much different, although perhaps somewhat better, since peak excursions would not necessarily have been engineered out.
>
> I have previously commented on the sorry state our nuclear-reactor industry would be in if radiation control had depended upon "threshold limit value" rather than engineering criteria. Can you imagine such regulation depending upon an army of inspectors with Geiger counters?

After describing some of the data developed in his studies of asbestos disease, Dr. Selikoff told Mrs. Strong that it was impossible to answer with any degree of accuracy the questions posed by her firm. He went on to point out that the weight of medical and scientific evidence concerning the occurrence of mesothelioma in nonoccupational circumstances, such as in families of asbestos workers and in people living in the vicinity of asbestos factories, bore heavily on the advisability of reaching a level of exposure as close to zero as possible. "The numerous instances of mesothelioma among workmen presumably exposed to asbestos as a result of indirect occupational exposure in shipyards, even in the absence of fiber counts thirty years ago, strongly points to asbestos disease at low levels of exposure," his letter continued. "Literally hundreds of cases of mesothelioma are now known to have occurred in such circumstances."

When I had finished reading the letter, I asked Dr. Selikoff why the Arthur D. Little company should be soliciting exposure-response judgments at this time.

"It is my understanding that A. D. Little has been awarded a contract by the Occupational Safety and Health Administration to formulate a consensus regarding exposure-response for asbestos disease," Dr. Selikoff replied.

"But the NIOSH criteria document and the Advisory Committee on the Asbestos Standard have already covered this ground by reviewing all the literature concerning asbestos disease," I said. "Not to mention the testimony given during four days of public hearings."

"True enough," Dr. Selikoff replied. "However, the A. D. Little people appear to have been specifically charged with determining the economic impact of the proposed permanent standard for occupational exposure to asbestos."

"Then why a questionnaire focused solely upon medical judgments?" I asked.

"That is a question I have been asking myself," Dr. Selikoff said dryly. "I don't know the answer. If you find out, please tell me."

During the next few days, I made dozens of telephone calls to people in various agencies of the Department of Health, Education, and Welfare, in the independent medical community, and in a number of labor unions, trying to ascertain what lay behind the involvement of Arthur D. Little, Inc., in the process of promulgating a permanent standard for occupational exposure to asbestos. The people I spoke with at NIOSH were clearly unhappy over the fact that a private consulting firm had been asked, in effect, to duplicate (if not amend) in the space of a few weeks all the effort that over a period of years had gone into the assessments, conclusions, and recommendations of the NIOSH document. "Look," one of them told me. "Our recommendation for a two-fiber standard and our conclusion that it is technically feasible were upheld by the Secretary of Labor's own Advisory Committee on the Asbestos Standard. However, A. D. Little is up to something that has no basis in science and no specific authorization in the Occupational Safety and Health Act. It's trying to form a consensus for what is sometimes called the 'socially acceptable risk' involved in occupational exposure to hazardous substances. In other words, it's trying to determine how much society is, or should be, willing to pay to avoid the loss of lives. The Act, however, clearly states that 'each employer shall furnish to each of his employees employment and a place of employment which are free from recognized hazards that are causing or are likely to cause death or serious physical harm to his employees.' "

When I called Sheldon Samuels, at the IUD, however, I was able to gain a new perspective on the matter. "The whole concept of economic-impact studies, as they now exist, began back in 1968 with the President's Task Force on Government Reorganization, which was headed by Roy L. Ash," Samuels told me. "The Ash commission called for an assessment of all government programs in terms of their effectiveness, and this has since been made by the Nixon Administration's Office of Management and Budget through a whole series of cost-benefit analyses that are conducted under the guise of environmental-impact studies. The present A. D. Little study has some extremely serious ramifications. Congress to the contrary, and throwing its Occupational Safety and Health Act to the winds, the executive branch of government has decided on its own that the cost to the employer of meeting any new occupational-health standard must fall within an economic range that is acceptable to industry. The major point, of course, is the government's order of priorities in this whole matter. I mean, how in the name of God can a serious, in-depth cost-benefit study of the proposed asbestos standard fail to assess as one of its first priorities the cost to the worker and the whole community of the terrible incidence of asbestos disease?"

When I asked Samuels if the Industrial Union Department had heard from the Arthur D. Little people, he told me that two representatives of the firm had visited him the previous week. "A Dr. Donald W. Meals and an engineer spent a whole day here," Samuels said. "They indicated that they had been brought into the picture to mediate between labor and industry, and to come up with a standard for occupational exposure to asbestos that would make everybody happy, and they asked for our help. During the past few days, I polled the members of our *ad hoc* Committee on the Asbestos Hazard, and we have decided to stand firm on the recommendations we made at the public hearings, and not to participate in the A. D. Little economic-impact study. We have good reasons for believing that the A. D. Little people were brought into the standard-setting process not just to satisfy the Office of Management and Budget but to justify the asbestos industry's position. We have learned, for example, that in the economic-feasibility part of their study the A. D. Little people are relying almost entirely on guess-estimates from the asbestos industry—particularly from the shipbuilding industry, in which the government has an enormous stake."

I asked Samuels if he was aware that the A. D. Little study was also seeking medical judgments on the incidence of asbestos disease.

"Indeed I am," he replied. "In fact, just the other day I heard that A. D. Little's so-called panel of medical experts is loaded with doctors who are or have been connected with the asbestos industry. It'll be interesting to see this roster when the final report of the study comes out."

In the second week of May, I visited Dr. Selikoff again and asked if he knew of any further developments in the involvement of Arthur D. Little in the standard-setting process. He told me that the firm had sent him the Phase II questionnaire of its economic-impact study. He also showed me a letter he had written on May 9th to Dr. Meals. The letter said, in part, "I have carefully considered the asbestos data forms sent me and am returning them to you unanswered. To have completed them, in my opinion, would only contribute further to an inappropriate exercise; my original misgivings (see my letter of April 3, 1972) are now amplified." In conclusion, Dr. Selikoff told Dr. Meals that the methodology upon which the A. D. Little study was based "could lead to serious misconceptions and misdirected advice."

The following morning, I telephoned Samuels to find out if he had any new information about the Arthur D. Little study, and he said he did.

"Have you looked at your mail today?" he asked.

I told him that I had not yet had time to do so.

"Well, see if there's a letter from me."

I went through the envelopes on my desk and saw that there was.

"Well, open it up and talk to me later," Samuels said. "You aren't going to believe what's inside."

After hanging up, I opened Samuels' letter and pulled out three documents that had been stapled together. The first was a press release for Monday, May 8th, sent out by the Connecticut Development Commission, in Hartford. The second was a letter written on May 4th by Mark Feinberg, managing director of the commission, to Jack Cawthorne, executive director of the National Association of State Development Agencies, in Washington. The third document was a letter written on Arthur D. Little stationery on January 25, 1972, by one John E. Kent. The letter from Feinberg to Cawthorne read:

> Dear Jack:
>
> We have learned that a Massachusetts based consulting firm, Arthur D. Little Inc., is attempting to sell a Connecticut manufacturer on moving its plant to Mexico. That information in itself is not startling, but what is startling is the fact that Arthur D. Little Inc. has a consulting contract from the U.S. Department of Labor to measure the impact of the standards being set for the asbestos industry under the recently enacted Occupational Safety and Health Act. And the company which Arthur D. Little is trying to move from Connecticut to Mexico is also in the asbestos industry. Thus it appears to me that at the same time as Arthur D. Little is carrying out a federal contract dealing with the asbestos industry and its problems, Arthur D. Little is also attempting to take one of our companies in that same industry to Mexico.
>
> This activity by Arthur D. Little in my opinion looks like a Trojan horse which I feel is highly improper. On the one hand, Arthur D. Little is accepting federal funds and on the other hand, it is attempting to help Mexico attract a firm directly involved in the federal project. Furthermore, it is shocking to me that a New England consulting company which has so frequently put forth the doctrine of helping

economic development here would "raid" a company in our state. As you know, we are certainly advocates of competition, free enterprise, and profit, but when a consultant presumably making a profit with federal dollars is at the same time attempting to take jobs away from our state and out of the country, it is a most serious matter.

I do not know what other companies in other states are being approached as our company was, and I feel strongly that the development directors of the other states should be warned about this Trojan-horse operation which certainly seems to be against the best interest of the people in the various states which may have similar situations. This operation by Arthur D. Little also seems to be contrary to all the efforts which we state development directors are making in cooperation with the U.S. Government to improve the national balance of payments and the economic development of our individual states.

Therefore, I am enclosing, for your use, the copy of the letter on Arthur D. Little stationery which was sent to the Connecticut company being "raided." I have taken out the company name and address in order to avoid embarrassment for the firm. I strongly urge you to send a bulletin to all our members alerting them to this serious problem as soon as possible.

Sincerely yours,
Mark Feinberg
Managing Director

P.S. You don't suppose there could be a relationship between the health and safety standards Arthur D. Little sets and the success of efforts to relocate American asbestos companies to Mexico?

After several phone calls, I learned that the corporation Arthur D. Little had attempted to relocate in Mexico was Raybestos-Manhattan, Inc., whose factory in Stratford, Connecticut, is a major producer of asbestos brake linings, clutch facings, and gaskets. A few weeks later, when I was able to examine a copy of Arthur D. Little's first report to the Department of Labor, which was entitled "Impact of Proposed O.S.H.A. Standard for Asbestos," I saw listed among its panel of experts John H. Marsh, who is the director of planning for Raybestos-Manhattan, and who had testified at the public hearings in Washington against the NIOSH recommendation requiring warning labels on asbestos products. Meanwhile, I had discovered that the asbestos industry was taking a hard look at the feasibility of moving some of its plants and facilities to Taiwan and Korea, where, presumably, it could operate unhindered by occupational-safety-and-health regulations. It was already becoming clear, however, that by involving Arthur D. Little, Inc., in the standard-setting process, the Department of Labor was attempting to counter the recommendations of the NIOSH criteria document, of the Secretary of Labor's Advisory Committee on the Asbestos Standard, and of the members of the independent medical and scientific community who had testified at the public hearings. It was also becoming clear how deeply the medical-industrial complex had succeeded in penetrating the workings of the government in matters relating to the prevention of industrial disease.

PART FOUR

No Tangible Effect on Sales and Earnings

While waiting to see what action the Secretary of Labor would take in the first week of June 1972, his deadline for setting a permanent standard for asbestos, I received a copy of "The President's Report on Occupational Safety and Health," which described what had been done to carry out the provisions of the Act during its first year of operation. The report, addressed to Congress, actually consisted of two separate reports, submitted to President Nixon in May by Secretary of Labor James Hodgson and by Secretary of Health, Education, and Welfare Elliot L. Richardson. Hodgson's report began by saying that each year, out of eighty million people employed in the civilian labor force, more than fourteen thousand are killed and two million two hundred thousand suffer disabling injuries on the job. He then said that there were no reliable figures on the number of employees who suffer minor, nondisabling injuries or become ill after being exposed to hazardous conditions. Toward the end of his report, however, Hodgson stated that occupational illnesses were "at least as great a problem as injuries" but that "it was much more difficult to develop a special program that would allow O.S.H.A. to effectively focus on them." He went on to say that, through its Target Health Hazards Program, the Occupational Safety and Health Administration would concentrate on five substances—asbestos, cotton dust, silica, lead, and carbon monoxide—which were among the most hazardous of more than eight thousand toxic substances currently identified by NIOSH.

Secretary Richardson was considerably more explicit in his assessment of the problem. His report matter-of-factly stated that according to recent estimates there were at least three hundred and ninety thousand new cases of disabling occupational disease in the United States each year. This figure was followed by one that boggles the mind: "Based on limited analysis of violent/non-violent mortality in several industries, there may be as many as 100,000 deaths per year from occupationally caused diseases." Richardson went on to say that at the end of 1971 NIOSH had completed a criteria document for the Secretary of Labor which recommended a two-fiber asbestos standard. According to Richardson, the two-fiber standard proposed by NIOSH would "protect against asbestosis and asbestos-induced [cancer]; be measurable by techniques that are valid, reproducible, and available to industry and official agencies; and be attainable with existing technology."

In appendixes to Richardson's report, there were long lists of contracts and grants that had been awarded by NIOSH to various universities, medical schools, corporations, and research institutes for studies relating to occupational safety and health. Among them were two grants and one contract, totaling more than a hundred and forty-six thousand dollars, that had been awarded to the Industrial Health Foundation, Inc., in Pittsburgh, and to Dr. Paul Gross, the director of the foundation's research laboratories, for studies relating to asbestos disease. As it happened, Dr. Gross had testified for Johns-Manville in workmen's-compensation cases, and the Industrial Health Foundation, Inc., was none other than the old Industrial Hygiene Foundation of America, Inc., the self-styled "association of industries for the advancement of healthful working conditions," which was hired by Pittsburgh Corning in the summer of 1963 to evaluate the asbestos-dust hazard at its plant, then newly acquired, in Tyler. Moreover, NIOSH's project officer for a contract under which more than fifty-eight thousand dollars had been supplied up to that time was Dr. Lewis J. Cralley, who, when director of NIOSH's Division of Epidemiology and Special Services, had ignored the data showing excessive asbestos-dust counts at the Tyler plant. One of the appendixes to Richardson's report also listed a contract for forty-eight thousand nine hundred and seventy-six dollars which had been awarded to Arthur D. Little, Inc., to "develop a priority rating system and identify general areas and specific problems where fruitful and necessary research in occupational safety should be undertaken."

In the middle of May, something occurred to shed light on the role of Johns-Manville—the world's largest producer and user of asbestos, with mines, mills, and some sixty manufacturing plants in the United States and Canada—in the public hearings in Washington, D.C., where, acting as *éminence grise* for the entire asbestos industry, it had mounted strenuous opposition to the proposed two-fiber standard. On May 18th, speaking before the annual meeting of the American Industrial Hygiene Association, in San Francisco, Dr. William J. Nicholson, of the Mount Sinai Environmental Sciences Laboratory, described the mortality study he, Dr. Irving Selikoff, and Dr. E. Cuyler Hammond had conducted which showed that a hundred and ninety-nine deaths had occurred during thirteen years—or sixty-five more than were to be expected according to the standard mortality tables—among six hundred and eighty-nine workers at Johns-Manville's plant in Manville, New Jersey. An examination showed that a vast majority of the excess deaths were the result of asbestos-induced disease. On May 23rd, Johns-Manville issued a press release quoting Wilbur L. Ruff, the manager of the Manville plant, as saying that the asbestos-dust levels that caused the fatal disease among the workers were those of past years, "when conditions were much worse than they are now." After praising the corporation's dedication to medical research and industrial hygiene, Ruff said that "though spending money doesn't mean a thing where human health is concerned, the six and a half million dollars we've spent on dust-control projects since 1949 does show that we're a company with conscience." Ruff also said that a recent dust survey conducted by the Occupational Safety and Health Administration indicated that the Manville plant had "an outstanding record in dust control."

This was the first I had heard of the Administration's inspection at Manville, so I called Robert Klinger, vice-president of Local 800 of the United Papermakers and Paperworkers Union, whom I had met at the public hearings in Washington, and asked him to tell me about it.

"Since April of 1971, when a government survey showed that some dust counts in the Manville textile operation were running as high as twenty fibers per cubic centimeter, there has been considerable improvement," Klinger said. "The 1971 survey and a previous survey that was conducted back in August of 1967 were buried in the files of the old Bureau of Occupational Safety and Health's Cincinnati offices, until Dr. Joseph Wagoner and Dr. William Johnson brought them to light in the summer of 1971, along with a 1969 medical survey showing that seventeen per cent of a hundred and seventy-nine workers in the Manville textile operation had X-rays that were consistent with asbestosis. When we learned about these hidden studies, last fall, we requested an immediate inspection of the Manville plant by the Occupational Safety and Health Administration. Its people came in November and December of that year, and returned in

April of this year. Their latest survey shows that eighty-one per cent of the dust stations in the plant are operating at between zero and two fibers per cubic centimeter of air; that seventeen per cent are operating at between two and five fibers; and that only two per cent indicate dust counts above the five-fiber level. This has come about simply because J-M has expended a great amount of money and effort over the past two years to engineer improved dust-control equipment and install it throughout the Manville plant."

I asked Klinger what had impelled the company to make this expenditure, and he told me that it was probably a combination of things. "We had a long and costly strike here in the autumn of 1970," he said. "As part of the settlement, the company guaranteed to make a real effort to reduce dust levels in the plant. Then, too, in 1969 Johns-Manville paid out nearly nine hundred thousand dollars in workmen's compensation in New Jersey for asbestosis alone, over and above what it may have settled out of court in litigation brought against it by workers, or families of workers, who had contracted asbestos-induced cancer. In addition, the work of men like Dr. Selikoff and Dr. Hammond was by then piling proof upon proof of the association between asbestos and disease. So the J-M people simply saw the writing on the wall, and decided they had better act."

At this point, I found myself remembering that without exception the testimony delivered by Johns-Manville officials and their medical associates at the public hearings in Washington had strongly urged the Secretary of Labor not to lower the standard for occupational exposure to asbestos from five to two fibers per cubic centimeter of air. Yet during the previous two years the company had undertaken to accomplish just that in the bulk of its operations in Manville, which has the largest complex of asbestos plants in the world. When I remarked upon this seeming contradiction to Klinger, however, he was not at all surprised.

"There's a simple explanation," he told me. "The Johns-Manville people sell huge amounts of raw asbestos fiber to competitors here and all over the world. In fact, they've pretty well got the chrysotile-asbestos market cornered. For example, according to their own announcement of 1970 earnings before taxes, the mining, milling, and selling of raw asbestos brought in twenty-five million dollars, which was nearly half of their total gross for that year. So, you see, a lower standard might drive the competitors to whom they sell raw fiber out of business, or cause them to look about for asbestos substitutes."

On June 6, 1972, Secretary Hodgson and George Guenther, who was the Assistant Secretary of Labor and the director of the Occupational Safety and Health Administration, announced the long-awaited decision on a permanent standard for asbestos. It served to confirm the doubts that had been expressed by Anthony Mazzocchi and other labor leaders about the Department of Labor's commitment to the provisions of the Occupational Safety and Health Act, for, despite the recommendations of both the Department of Health, Education, and Welfare's National Institute for Occupational Safety and Health and the members of the Secretary's own Advisory Committee on the Asbestos Standard, the new regulations stipulated that the five-fiber standard would remain in effect for four more years, and that a two-fiber level would become effective only on July 1, 1976.

Reaction to the ruling was immediate, and it came from all quarters. On June 7th, John Jobe, vice-president in charge of operations for Johns-Manville, issued a statement from the company's headquarters, in Denver, assuring stockholders that the new controls would have "no tangible effect on sales and earnings." On the same day, *The New York Times* ran a short article on its inside back page that said, "The Occupational Safety and Health Administration ordered today a continuation of asbestos dust exposure limits for four more years despite the recommendations by a scientific panel that they be cut by more than half." *The Wall Street Journal*, in an article in its June 7th issue, put a slightly different interpretation on the news. "The Occupational Safety and Health Administration announced tough new curbs on asbestos in plants, but gave employers four years to comply," it stated. That same day, Sheldon Samuels, of the Industrial Union Department, told me that, in addition to giving industry two extra years in which to comply, the Administration had rejected recommendations from NIOSH for medical surveillance of asbestos workers, for medical record-keeping, and for labels on asbestos products warning that inhalation of asbestos could cause asbestosis or cancer. "Moreover, because the Administration has only a handful of industrial hygienists, the determination of actual levels of asbestos dust in workplaces will depend upon tests conducted by the employers," Samuels said. "No controls will be required if an employer finds, or believes, that dust levels do not exceed the standard, which, of course, makes a monstrous joke of the whole business. In fact, the new standard is so appallingly deficient that we plan to fight it in the courts."

Dr. Selikoff was quoted by one source as saying, when he was asked his opinion of the new standard, that his remarks "would have to be written on asbestos-coated paper." Then, in a speech he gave in Washington on June 12th, which was quoted in an article in the *Times*, he predicted that tens of thousands of workers exposed to asbestos would die because of inadequate regulations issued by the Occupational Safety and Health Administration, and added that if the Administration showed the same disregard for essential precautions in setting standards for other toxic substances "we face an unparalleled disaster to the working people in our country." He charged Assistant Secretary Guenther with creating a situation in which "workers exposed to asbestos in any trade are required to work under conditions which permit them to inhale twenty million or even thirty million fibers in a working day."

In the same article, Guenther was quoted as saying that Dr. Selikoff had chosen to overdramatize the matter. "There is no question that exposure to asbestos is most hazardous," Guenther declared. "We believe that the new standards will provide substantial and real protection for exposed workers, and that they provide for reductions in levels of asbestos exposure that are reasonable and, in our judgment, based on tests from many quarters." Guenther did not identify the tests and the many quarters from which they supposedly came, but he certainly could not have been referring to any tests described in a story entitled "Asbestos: Airborne Danger," which appeared in the May–June, 1972, issue of *Safety Standards*, the official bi-monthly magazine of his own Occupational Safety and Health Administration. After stating that "asbestos has been recognized as one of the most hazardous of air contaminants," the article described only one test concerning the effects of asbestos exposure. That was the study of the disastrous mortality experience of the asbestos-insulation workers which had been conducted by Dr. Selikoff and Dr. Hammond.

For sheer irony, however, there was an event in the early part of June that rivaled Guenther's contradiction of his own house publication. On June 9th, NIOSH held its first-anniversary celebration and awards ceremony, at the Cincinnati convention Center. Among the recipients of awards was Dr. Cralley, who was presented with the Public Health Service Meritorious Service Medal, "in recognition of his research into developing safe worker exposure levels to such potential occupational hazards as uranium, asbestos, silica, beryllium, and diatomaceous earth dust."

Up to that time, no one knew what conclusions the Arthur D. Little people had drawn in their study of the proposed asbestos standard, because the firm did not receive permission from the Department of Labor to print and distribute a report of the study until June 8th. Within a few days, however, Dr. Nicholson received his copy of the report and a letter from Dr. Donald Meals, of Arthur D. Little, Inc., thanking him for "your support as a member of one of the panels of experts." One indication that the Arthur D. Little people must have put the report together very hastily was that Dr. Nicholson, who had contributed to the study, was not listed in the report as a member of the panel of health experts, while Dr. Selikoff, who had written two letters to the Arthur D. Little people telling them that the methodology of their study had little scientific validity, did not receive a copy of the report or a letter of appreciation from Dr. Meals for his contribution to the study but was listed in the report as a member of the panel of health experts.

No one seems to know what the Arthur D. Little people had in mind in all this, but the possibility that they felt the need to achieve some semblance of balance and impartiality in their panel of health experts presents itself to anyone examining the roster of eleven men listed as its members. In the order in which their names appeared, they were Dr. Edward A. Gaensler, professor of surgery and director of thoracic services at Boston University's Medical Center, who has made useful contributions to the study of asbestos disease, and who has also been retained by Johns-Manville to examine workers at its asbestos-wallboard plant in Billerica, Massachusetts; Dr. Thomas Davison, medical director of Johns-Manville; Dr. John McDonald, the chairman of the Department of Epidemiology and Health of McGill University, in Montreal, and the author of a study entitled "Mortality in the Chrysotile Asbestos Mines and Mills of Quebec," which was financed by the Quebec Asbestos Mining Association, of which Johns-Manville is a leading member; Dr. Cralley, the former director of the Division of Epidemiology and Special Services; in whose files Dr. Johnson and Dr. Wagoner had uncovered hidden data showing grossly excessive levels of asbestos dust in the Tyler plant and other asbestos factories across the land as well as data showing an appalling rate of deaths from asbestos disease among asbestos-textile workers; Howard E. Ayer, the former assistant director of the division, who, as its senior industrial hygienist, was involved in the interpretation of the dust levels measured over the years at asbestos factories; Dr. George Wright, head of medical research at St. Luke's Hospital in Cleveland, and a longtime paid consultant of Johns-Manville, who testified for the corporation at the public hearings in Washington; Dr. Hans Weill, a professor of medicine in the pulmonary-diseases section of the Tulane University School of Medicine, who had been given financial support by the Quebec Asbestos Mining Association to conduct a study of asbestosis among men working at a Johns-Manville cement-products plant at Marrero, Louisiana; Dr. W. Clark Cooper, a former head of the old Bureau of Occupational Safety and Health and a partner in Tabershaw-Cooper Associates, Inc., a consulting firm in Berkeley, California, which has done research contract work not only for the National Insulation Manufacturers Association, of which Johns-Manville is a member, but also for Pittsburgh Corning; Dr. Philip E. Enterline, a professor in the Department of Biostatistics of the University of Pittsburgh, who was retained by Johns-Manville to study the health experiences of its retired employees; Commander Samuel H. Barboo, an industrial hygienist in the Medical Service Corps of the United States Navy (a large purchaser of asbestos insulation), who had never been involved in any studies concerning the health effects of asbestos; and, lastly, Dr. Selikoff.

Considering the backgrounds of most of the members of the expert health panel, and the fact that two additional panels of experts consisted of a committee of thirteen men representing private shipbuilding companies and a group of twelve men from various asbestos-producing companies (including two executives of Johns-Manville), it was hardly surprising that the Arthur D. Little people concluded in their report that "reduction of the exposure of workers to asbestos dust from present levels to five fibers per cubic centimeter will significantly reduce asbestos-related diseases and achieve more than 99% of the benefits attainable from the control of dust levels." (In talking about a reduction "from present levels," they apparently forgot that an emergency five-fiber standard had been in effect for nearly five months.) Their report went on to say that because of the cost the two-fiber standard very probably could not be met by the asbestos industry within the two-year period recommended by NIOSH; that the standard could not be achieved at any cost within two years by companies engaged in on-board ship repair in private shipyards (asbestos insulation being used extensively in shipbuilding); and that it would certainly lead to intensified foreign competition in this field and to the imposition of "difficult problems and costly solutions" upon United States Navy shipyards. In making this statement, the Arthur D. Little people overlooked evidence and statistical tables in the NIOSH criteria document showing that many asbestos plants were already operating at or below a two-fiber level, and that most of the asbestos industry could comply with a two-fiber standard without undue technical hardship. And, piling oversight upon oversight, they went on to say that estimates of when each segment of the asbestos industry could meet various fiber levels "show at a first glance that only a twelve-fiber standard can be met immediately"—an assertion that ignored the fact that a twelve-fiber standard was supposed to have been in effect from 1968 until the temporary emergency standard of five fibers was declared by the Secretary of Labor six months earlier.

Later in their report, in a section that was entitled "Benefits from Asbestos Exposure Control Standards," the Arthur D. Little people got around to defining what they meant by benefits:

> The case against asbestos dust is firm and unquestioned by those familiar with available research data. Selikoff and others have amply demonstrated an association between exposure to asbestos fibers and increased morbidity. It follows that reduction of the hazard will provide increased freedom from disease and longer life for those working with or near asbestos.
>
> The question of how closely the goal of zero risk to asbestos-related diseases can be approached requires further exploration. The removal of this hazard requires changes that inevitably involve the expenditure of time and money, and in the world of business (including working men when jobs are at stake) the relationship between

> benefits and costs is an important issue. Numerous examples of the refusal of people individually and collectively to pay even modest inconvenience costs to completely remove risks demonstrate that eliminating a hazard at any cost is not always feasible. However, those who must pay the price for removing the hazard may be willing or able to do so within limits. Finding these limits usually involves comparing less than the maximum obtainable benefits with associated costs. This, of course, is the familiar cost/benefit framework for evaluating alternative courses of action. While we do not believe a purely quantitative cost/benefit analysis is feasible or desirable here, the conceptual scheme is useful. In the next section, we examine the probable costs for various reductions (fiber levels) in the hazard of asbestos dust. To place these costs in a perspective that may be useful in setting policy, it is important to estimate the benefits associated with each of several such levels of risk.

At this point, the Arthur D. Little people drew attention to Table 2 of their report, which incorporated estimates from eight of the eleven listed health-panel members as to the incidence of asbestosis, lung cancer, and mesothelioma among a hundred workers exposed to various levels of asbestos fibers during an eight-hour working day over a period of forty years. A footnote stated that by the time the report went to press two additional responses had been received, which did not change either the median or the range. The footnote did not mention that Dr. Selikoff's responses to this questionnaire were the only ones not included. As a result, Table 2 concluded that not one worker in a hundred would develop asbestosis after being exposed to two fibers per cubic centimeter for forty years, whereas Dr. Selikoff had estimated that fifty-five of a hundred workers would develop the disease under these conditions. In addition, Table 2 concluded that only one worker in a hundred would develop asbestosis after working for forty years in an environment containing five fibers per cubic centimeter, whereas Dr. Selikoff had estimated that eighty-five workers out of a hundred would contract the disease under such conditions. As for mesothelioma, the Arthur D. Little report concluded that only one out of a thousand workers would be afflicted with the disease after forty years of exposure to two fibers per cubic centimeter, whereas Dr. Selikoff had estimated that four out of every hundred workers would develop mesothelioma under these circumstances. And Table 2 concluded that only two out of a thousand workers would develop mesothelioma after forty years of exposure to a working environment containing five fibers per cubic centimeter of air, whereas Dr. Selikoff had estimated that seven out of a hundred workers would be afflicted with mesothelioma at this level.

Now, using Table 2 as a springboard, the Arthur D. Little people leaped to other unfounded conclusions. "It is apparent from this set of judgments that relatively large benefits correspond to the reduction of exposure from thirty to twelve fibers, and from twelve to five fibers," they wrote. "The judgments suggest, however, that a further reduction of the exposure level to two fibers is attended by very small benefits—on the order of less than one per cent." They could make such an assertion, of course, only because they had not seen fit to include in Table 2 the responses of Dr. Selikoff, who is widely regarded as one of the world's foremost epidemiologists in the field of asbestos disease, and whose epidemiological investigations of asbestos disease had never been supported by any segment of the asbestos industry. Having made the assertion, however, the Arthur D. Little people bounced along to others:

> Data on bronchogenic cancer and mesothelioma suggest that these diseases are also related to the degree of exposure. The numbers are small, however, and experience more limited than that available for asbestosis. The only inference we are prepared to draw from these data is that their explicit consideration by the panel members very probably yields better estimates than if they had not been included. Continuing studies of these diseases among asbestos workers should contribute to estimates that better justify interpretation and speculation than these.

All this, of course, overlooked the fact that in their studies of the disastrous mortality experience of six hundred and thirty-two asbestos-insulation workers in New York City and Newark, and of nine hundred and thirty-three men who had worked at the asbestos-insulation factory in Paterson, New Jersey, Dr. Selikoff and Dr. Hammond had furnished incontrovertible proof that two hundred and thirty out of three hundred and nineteen excess deaths among these men were caused by some form of cancer.

The rest of the Arthur D. Little report consisted of analyses of such things as the gross-sales profits of various segments of the asbestos industry and the estimated economic impact of various asbestos standards on shipbuilding companies and on manufacturers of asbestos products. A section entitled "Economic Impact on Manufacturers" contained this passage:

> With regard to technical feasibility, we judge that the five-fiber level is achievable. Even with the best available techniques, however, we do not know whether a two-fiber limit could be met. (In fact, we cannot be completely certain about the five-fiber limit until the best available equipment has been installed and evaluated.) Thus a reliable assessment of the validity of the "guesstimates" put forth on the cost of compliance to the two-fiber level is not really possible until technical feasibility has been established. In the meantime, the estimates shown in Table Six are the best that have been developed.

The estimates, or "guesstimates," in Table 6 may well have been the best that were available, but to me they were incomprehensible, so I went to Mount Sinai and asked Dr. Nicholson, whose copy of the report I had been reading, to interpret them for me.

Dr. Nicholson shook his head and gave a weary smile. "The cost estimates obtained by Arthur D. Little from its panel of experts are inappropriate on two counts," he said. "First, the panelists were asked to estimate time and costs to achieve specific dust levels, and not costs to achieve effective worker protection. Second, the cost estimates were obtained by soliciting guesses from representatives of the asbestos industry rather than by reviewing the cost and effectiveness of existing installations. The whole approach is like the old story of asking the fox to guard the chicken coop. Here the fox has been asked how many chickens he would kill. 'Why, hardly any,' he replies. Then the fox is asked how much it would cost to keep him from killing just those few chickens. 'Oh, much too much to consider,' he answers."

Glancing over the names and corporate affiliations of Arthur D. Little's twelve-man expert panel from the asbestos industry, I saw that all eleven of the companies they represented had sent officials to the public hearings to testify against the proposed two-fiber standard. When I pointed this out to Dr. Nicholson, he shook his head again.

"The ins and outs of this whole affair constitute an endless labyrinth that never ceases to amaze me," Dr. Nicholson said. "I just noticed, for example, that both the Certain-teed Products Corporation and Nicolet Industries, Inc., are represented on the panel, and that reminds me of a story I read in *The Wall Street Journal* earlier this month. It was about the old Keasbey & Mattison Company, which used to make milk of magnesia in the town of Ambler, Pennsylvania, near Valley Forge. The article described an incident in the early history of the company, when Dr. Royal Mattison accidentally spilled some milk of magnesia on a hot pipe and found that it adhered. That brought the Keasbey & Mattison people into the insulation business. In 1962, however, the company went out of business and sold its facilities to Certain-teed Products, which started manufacturing asbestos-cement pipe. Since that time, Ambler has also been the site of a plant operated by Nicolet Industries, which manufacturers other asbestos-cement products. In the *Journal* article, there was a description of a large open-air dump that has existed in the town since 1867, that is owned by Nicolet and Certain-teed, and that is still used by Certain-teed, which adds twenty-seven hundred tons of crushed asbestos pipe to the dump each year. Naturally, this aroused my curiosity, so I drove down to Ambler the other day to take a look at it. When I got there, I could hardly believe my eyes. The dump not only snakes diagonally through the very center of the town, which has a population of about eight thousand, but it is fifty feet high, anywhere from one to two city blocks wide, and about ten city blocks long. In fact, it is estimated to contain some million and a half cubic yards of waste material. I brought back a dozen or so samples of debris to our mineralogy laboratory for analysis, and we found that all of them contained large amounts of chrysotile-asbestos fiber. The incredible thing, however, is that while the townspeople of Ambler want to get rid of the dump—it's an eyesore, of course—almost no one down there seems to be aware of the health hazard it poses. Kids play on an asphalt basketball court that has been built smack on top of material from the dump, and is literally covered with loose asbestos fibre and wads of waste material containing asbestos. Not only that but the dump itself is pockmarked with holes and tunnels dug over the years by kids searching for old milk-of-magnesia bottles, which have become collector's items. As you may already be aware, cases of mesothelioma have been reported among people whose exposure to asbestos was that as children they had played on asbestos dumps."

A few days later, I was reminded of the Ambler dump while looking at an exhibit submitted as evidence during the public hearings by Bruce J. Phillips, a senior vice-president of Certain-teed. "We do not feel that there is sufficient medical justification for a two-fiber limit at this time," Phillips had said. "We propose a five-fiber standard." However, this statement of Phillips's aroused my curiosity less than the next. "We feel that asbestos scrap and waste, including asbestos dust, can be disposed of in quantity only in a landfill, where the waste can be covered each day, and will present no danger to anyone. Furthermore, these disposal problems are solid-waste-management problems to be covered by the Environmental Protection Agency, and not OSHA," he said. Whether Phillips had the Ambler dump in mind when he made this statement is a matter of conjecture, but it is a matter of record that the EPA had not then got around to declaring a standard for asbestos dust in the ambient air.

Toward the end of June, Herman Yandle, the former union committee chairman at the Tyler plant, called me from Hawkins, Texas, to inform me that ten thousand-odd bags of asbestos fiber left in the warehouse after Pittsburgh Corning closed the factory, and either buried or otherwise disposed of virtually all the rest of its innards, had been shipped to Canada during the last part of May. Yandle had been told that the fiber had been bought by a company with facilities on Manitoulin Island, in Ontario.

That piece of information sent me to a map, where I discovered that Manitoulin is a very large island in the northern part of Lake Huron, about a hundred miles east of Sault Ste. Marie. As a result of inquiries I made to find someone who might be able to tell me something about the company, I got in touch with Dr. Ernest Mastromatteo, who is the director of the Environmental Health Services Branch of the Ontario Ministry of Health, in Toronto.

"Manitoulin Island is way up in the wilds, and I've never heard of any asbestos plant up there," Dr. Mastromatteo said. "But I'll have our people look into it, and get back to you as soon as we have something to report."

During July of 1972, another intricate tier was added to the labyrinth constructed by the medical-industrial complex in its dealings with the problem of occupational exposure to asbestos, and I heard about it from Dr. Selikoff. I had not seen him since the Department of Labor decided to keep the five-fiber standard for four more years, and I had gone to Mount Sinai to ask him about an article from England which, I had been told, bore on the question of the asbestos standard.

"As you know, the two-fiber standard we had hoped for was designed only for the prevention of asbestosis, and not of cancer," Dr. Selikoff said. "It was first proposed back in 1968 by the British Occupational Hygiene Society's Committee on Hygiene Standards. At that time, the committee reported in the *Annals of Occupational Hygiene*, a respected British medical journal, what appeared to be strong evidence suggesting that a person working with chrysotile asbestos could be exposed for fifty years to a level of two fibers per cubic centimeter with practically no risk of developing asbestosis. The committee therefore recommended a two-fiber level for occupational exposure to asbestos—a proposal I considered to be reasonable, and one which was adopted by the British Inspectorate of Factories that same year. The British action was one of the chief determining factors in the decision of NIOSH to recommend a two-fiber standard to the Secretary of Labor. The report of the society's committee noted that its recommendation was based entirely upon information provided by two of the committee's members who were employees of the Turner Brothers Asbestos Company, Ltd., of Rochdale, England—one of the largest asbestos companies in Britain. They were Dr. John F. Knox, who was then chief medical officer for Turner Brothers, and Dr. Stephen Holmes, the company's industrial hygienist. Dust counts had been made at the Turner Brothers factory in Rochdale for many years, and in 1966 Dr. Knox and Dr. Holmes had undertaken to correlate the health status of current employees with past exposure levels. Accordingly, they had reviewed chest X-rays of two hundred and ninety workers at the factory who had been employed for ten years or more after January 1, 1933, and were still employed there as of June 30, 1966. They then reported to the committee that among these employees they had found only eight whose X-rays could be diagnosed as asbestotic. Among the eighty-one men exposed to levels of somewhat more than ten fibers per cubic centimeter for twenty years or more, they found only six with relevant X-ray changes, and out of thirty-seven workers exposed over a twenty-to-thirty-year period to levels of about four fibers per cubic centimeter, they found only one man with a possibly asbestotic X-ray abnormality. This was comforting information indeed, and the committee, after providing a safety margin by halving the four-fiber level, recommended the two-fiber standard in the full assurance that it would prevent the occurrence of asbestosis."

I was reminded by Dr. Selikoff's remarks that Dr. Holmes had flown to the United States in March to testify in behalf of the American asbestos industry at the Department of Labor's public hearings, and had stated his opinion that workers could safely inhale air containing four or five fibers per cubic centimeter. When I mentioned this to Dr. Selikoff, he told me that not long after the hearings ended, the April 1972 issue of the *Royal Society of Health Journal* arrived in the United States. "It contained an article by Dr. Hilton C. Lewinsohn, a young South African physician, who had replaced Dr. Knox as chief medical officer of Turner Brothers," Dr. Selikoff said. "Dr. Lewinsohn wrote that in December of 1970 he had given chest X-rays to workers employed at the Turner Brothers Rochdale factory. Although some workers had left the firm or had died, and although others had only recently completed ten years of employment, the men with long-term exposure who were examined by Dr. Lewinsohn were in large part the same men who had been studied four and a half years before by Dr. Knox and Dr. Holmes. However, Dr. Lewinsohn's assessment of X-ray findings among them was quite different from that of his predecessor, Dr. Knox. He reported that more than half of those men X-rayed twenty years or longer after first exposure to asbestos showed some abnormal lung changes. Moreover, upon analyzing these changes he determined that almost forty per cent of the men who had been employed at the factory for twenty years or more had lung scarring consistent with asbestosis. Unfortunately, our hearings were concluded by the time the *Royal Society of Health Journal* arrived, so there was no opportunity to ask Dr. Holmes about the apparent gross discrepancy between the two sets of findings."

I asked Dr. Selikoff how he explained the startling discrepancy, and he replied that it must lie in the interpretation of the X-rays, since it was highly unlikely that there could have been such a marked increase in detectable lung disease in four and a half years. He added that Dr. Nicholson had prepared a statistical analysis of the dose-disease responses that could be derived from these two conflicting sets of data, and had determined that they showed as much as a tenfold difference in the incidence of X-ray changes characteristic of asbestosis among the workers at the Turner Brothers Rochdale factory.

"Do you mean that Dr. Lewinsohn found ten times as much disease among some Turner Brothers workers as Dr. Knox and Dr. Holmes had found four and a half years earlier?" I asked.

"Yes," Dr. Selikoff replied. "At least, that is what the data published so far suggest."

"So the two-fiber standard, even if it had been adopted without the four-year delay, is not sufficient," I said.

"Until the discrepancy is resolved, it would appear useful only as an interim measure for the prevention of asbestosis," Dr. Selikoff replied. "Further, one should remember that when the British Occupational Hygiene Society recommended the two-fiber standard, it took the prudent position that the standard was intended only for the prevention of lung scarring, since it was not possible at that time to specify a concentration of asbestos in the air which was known to be free of the risk of inducing cancer."

The possibility presented by this new data—that industry influence might have had an effect on medical considerations concerning the problem of occupational exposure to asbestos in England—came as a surprise to me for two reasons. The first was that until Dr. Selikoff and Dr. Hammond had conducted their pioneering studies of the asbestos-insulation workers, in the early nineteen-sixties, by far the best and most thorough investigation of asbestos disease had been conducted in England, where asbestosis had been recognized as a serious occupational-health hazard since the nineteen-twenties. The second was that I had talked at length with several leading English epidemiologists in London some months earlier, and when I told them that occupational-health data were being suppressed and ignored in the United States, they had solemnly assured me that, because of the independent character of English medicine and of the British government's occupational-health agencies, such a situation could not exist in their country.

When I told Sheldon Samuels the news from England, he was not at all surprised. "Business is always business," he said. "In fact, I have a new wrinkle for you here at home. When I came to the Industrial Union Department last year, I undertook an investigation of the company-doctor system, because it occurred to me that the system was operating to thwart the Occupational Safety and Health Act. A month or so after I began the investigation, I talked with Dr. Norbert Roberts, who is associate medical director of the Standard Oil Corporation of New Jersey [now the Exxon Corporation], and was then president of the Industrial Medical Association. Dr. Roberts told me that he and his associates in the Industrial Medical Association hoped to reform and improve the industrial medical profession by setting standards for the professional performance of company doctors. They proposed to expand the activities of an organization called the Occupational Health Institute by creating a new program to validate occupational-health programs and to set up standards that would enable industry, through voluntary compliance, to clean up the major health problems afflicting workers in the United States. The Occupational Health Institute is affiliated with the Industrial Medical Association, and, according to Dr. Roberts, the new program would derive its chief support from the Industrial Medical Association, the American Industrial Hygiene Association, and the American Association of Industrial Nurses, Inc.—all of which are organizations supported and controlled by industry. After listening to him, I told Dr. Roberts that in my opinion voluntary ethical standards would not accomplish the purpose of raising the professional performance of company doctors but should be part of enforceable OSHA regulations in order to guarantee the participatory role of workers envisioned in the Occupational Safety and Health Act, such as their guaranteed right to have access to medical records and to records pertaining to exposure to toxic substances. That was the last I heard of the matter for some time. In November of 1971, however, I sent Dr. Roberts the results of the NIOSH survey of Pittsburgh Corning's Tyler plant and asked him to have the Industrial Medical Association conduct its own investigation of the situation. Then, in February of this year, I sent him an account of the charges Mazzocchi had leveled against Pittsburgh Corning at his press conference, which included some information about how the affair had been handled by the company's medical consultant, Dr. Lee B. Grant. In the meantime, Dr. Roberts and his associates had pursued their plan to set up the new program of the Occupational Health Institute. At their request, a meeting to discuss the aims of the institute was held here at the IUD on March 24th. In addition to Dr. Roberts and myself, the meeting was attended by Dr. Duane Block, medical director of the Ford Motor Company; by Dr. Gilbert H. Collings, general medical director of the New York Telephone Company, who had been named to head the Occupational Health Institute; and by Dr. Marcus Key, the director of NIOSH, who was being asked to give NIOSH support to the institute's new program. According to Dr. Roberts and Dr. Collings, an accreditation commission initiated by the institute would pressure management to become enlightened in the field of occupational health by certifying valid industrial medical programs. Imagine my surprise, however, when I learned that Dr. Roberts and his colleagues were proposing Dr. Grant as one of the members of this accreditation commission. I pointed out that the very fact that Dr. Grant was being considered for such a position while serious allegations concerning his professional conduct remained to be resolved scarcely inspired confidence in the purpose and viability of the Occupational Health Institute's new program. Subsequently, Dr. Roberts and his associates must have brought pressure to bear upon Dr. Grant to remove himself from consideration, for early this month I received a telephone call from William D. Kelley, the director of NIOSH's Division of Training, in Cincinnati, telling me that he had received a letter from Dr. Grant complaining that, because of his consultant relationship with Pittsburgh Corning, I considered him to be anti-labor and, therefore, unacceptable as a commissioner. Now, what do you think of that?"

I told Samuels that I found it surprising, but I was really thinking that the medical-industrial complex went about its business in ways whose intricacies were a wonder to behold, and that the story I had been following for so many months seemed not only to repeat itself endlessly but to employ the same cast of characters.

"And now for the clincher," Samuels said. "Guess what outfit has just received a seventy-one-thousand-four-hundred-and-eighty-one-dollar contract from NIOSH, in order to—and I quote—'develop and validate criteria for program performance standards of occupational-health programs which will provide guidelines to NIOSH in promoting the development of such programs as well as provide guidance to facilities in establishing and/or upgrading their own operational program standards.' "

"The Occupational Health Institute?" I said.

"Of Chicago, Illinois," Samuels replied.

A few days later, I telephoned Dr. Mastromatteo in Toronto, and asked him if he had been able to find out anything about an asbestos plant on Manitoulin Island.

"I'm sorry to say I haven't," Dr. Mastromatteo replied. "We've checked all the available records and made inquiries by phone, but there doesn't appear to be any asbestos industry at all on Manitoulin Island."

I apologized to Dr. Mastromatteo for bothering him with what was obviously a false lead, and thanked him for taking the trouble to follow it up. Then I telephoned Herman Yandle in Hawkins, and told him what had happened.

"Well, that's what they told the boys at the plant when the stuff got sent out," Yandle said, with a chuckle. "But I got a new address for you just the other day, from a fellow who saw a shipping tag. The asbestos went to Canada, all right, but not to Manitoulin Island. It got sent to a company called Holmes Insulation, Ltd., at Point Edward, Ontario."

I called Dr. Mastromatteo back and told him what I had learned from Yandle.

"We'll try to run it down for you," he said. "I've never heard of Holmes Insulation, but I know where Point Edward is. It's about two hundred miles west of here, in the middle of a large petroleum-chemical complex."

In the second week of September, I received some rather interesting mail. From the August issue of the *Archives of Environmental Health* I learned that Dr. Grant had been elected treasurer of the American Academy of Occupational Medicine. In a clipping from the August 25th edition of the Cincinnati *Post*, I read that Dr. Wagoner and Dr. Johnson had been looking into the spraying of asbestos insulation on the steel girders of a new Procter & Gamble technical center in Blue Ash, Ohio, a suburb of Cincinnati. The *Post* article stated that the spraying of asbestos insulation in construction had been banned in New York, Boston, Philadelphia, and Chicago, but that in Cincinnati and the rest of Ohio there were no ordinances either banning or controlling its use. The article went on to quote Dr. Mitchell R. Zavon, Assistant Health Commissioner for Cincinnati, who said that he had not seen fit to issue any regulations, since the evidence incriminating asbestos as a health hazard "is not sufficiently clear-cut." This surprised me, since I knew that Dr. Zavon was a member of the sixteen-man Threshold Limit Value Airborne Contaminants Committee of the American Conference of Governmental Industrial Hygienists, so I filed the newspaper clipping away with a mental note to make some inquiries about him.

A week or so later, Samuels sent me a copy of a statement

delivered before the House Select Committee on Labor by Jacob Clayman, administrative director of the Industrial Union Department, concerning the activities of the government in enforcing the Occupational Safety and Health Act. It was obvious that Clayman did not think much of the government's performance. After reminding the committee members that the Department of Health, Education, and Welfare had estimated that there might be as many as a hundred thousand deaths each year from occupationally caused disease, he told them that seventeen months after the Act went into effect there were only four hundred inspectors to enforce its provisions in more than four million workshops, and that only one new health standard—that for asbestos—had been set during that time. "Even then, we have had to complain about its inadequacies," Clayman said. "Indeed, we have gone to court to emphasize its failure, in our judgment, to pursue the basic purpose of the law." Clayman said that his organization had never ceased to point out that not one of the nine carcinogens rated by the Conference of Hygienists and by forty-five states as being too dangerous for any exposure at all was included in the federal standards. "Benzidine is one of these," he declared. "More than seventeen hundred tons of this chemical and three of its most commonly used compounds are produced, distributed, and used in this country each year. We have no idea how many workers are exposed. We have no idea as to how many of the hundred thousand deaths that HEW reports are due to exposure to this and other carcinogens."

Up to this time—the late summer of 1972—my investigation of the workings of the medical-industrial complex had evolved out of the critical situation that had existed at Pittsburgh Corning's Tyler plant, and had focused on the problem of exposure to asbestos. During the previous six months, however, I had been told by Mazzocchi, Samuels, and others that the complex was hard at work trying to put the lid on dozens of occupational-health problems involving, among other things, beryllium, benzidine, and beta-naphthylamine. (Indeed, Samuels and the Industrial Union Department had been pressing unsuccessfully for the Occupational Safety and Health Administration to take action on known chemical carcinogens for more than a year.) Now, reminded by Clayman's testimony of the multiplicity of the hazards, I decided to take a closer look at some of them.

On Monday, September 25th, I caught an early-morning flight to Cincinnati and went to the Division of Field Studies and Clinical Investigations to ask Dr. Johnson, whom I had not seen since March, what was being done about the problem of occupational exposure to beryllium, benzidine, and beta-naphthylamine. "Let's start with beryllium," Dr. Johnson said. "It is a metal that has been used in the United States since the nineteen-twenties, when it was alloyed with copper and other metals in order to give them greater tensile strength and fatigue resistance. During the late nineteen-thirties, it was used extensively in the manufacture of fluorescent-lamp tubes. Subsequently, however, thanks to the pioneering work of Dr. Harriet L. Hardy, who was then an occupational-health physician at the Massachusetts Institute of Technology, beryllium was found to be a cause of acute and chronic pulmonary disease among workers in the fluorescent-lamp industry, and in 1949 its use in the manufacture of those appliances was discontinued. Because of its light weight, its rigidity, and its stability, beryllium has achieved wide use in other areas. It is employed in nuclear reactors, in aerospace structural materials and inertial guidance systems, and in satellite antennae, rocket-motor parts, heat shields, rotor blades, and airplane brakes. It also appears to have considerable potential as a solid rocket fuel. Suffice it to say that beryllium and its compounds have an important use in modern technology, and that, as I've indicated, exposure to them can result in beryllium disease, which can cause death from pulmonary insufficiency, or right-sided heart failure."

Dr. Johnson went on to say that during the summer of 1971, while he was unearthing from his predecessors' files the data on the Tyler plant and other asbestos factories, he had come across the report of an environmental survey that had been conducted in 1968 by engineers from Dr. Cralley's Division of Epidemiology and Special Services at a factory owned by Kawecki Berylco Industries, Inc., in Hazleton, Pennsylvania. "As in the case of the Tyler studies, no action had been instituted as a result of the Hazleton-plant survey," Dr. Johnson told me. "Yet the data showed incredibly high airborne levels of beryllium dust, especially in the factory's attrition-mill operations, where beryllium powder is made. Indeed, the engineers had recorded concentrations in the Hazleton plant as high as two hundred micrograms of beryllium per cubic meter of air, though the Conference of Hygienists had already adopted the Atomic Energy Commission's standard of a peak value of twenty-five micrograms per cubic meter of air and a time-weighted average of only two micrograms per cubic meter."

When I asked Dr. Johnson what he had done with the information he uncovered, he told me that, just as in the case of the data on conditions in the Tyler plant—data that were collected and ignored—he had relayed them to Anthony Mazzocchi and Steven Wodka, of the Oil, Chemical and Atomic Workers International Union, which, as it happened, also represented workers at the Hazleton plant. "As a result, just as in the case of Tyler, the union expressed concern to our division, and we conducted a comprehensive environmental survey of the plant in November of 1971," Dr. Johnson said. "The beryllium-dust levels we measured at that time showed that the company's powder-handling operations were still grossly out of control. In March of 1972, beryllium levels at the Hazleton plant were measured again, by the Pennsylvania State Department of Health, and once more the levels far exceeded the recommended threshold limit value of two micrograms per cubic meter of air. Consequently, on July 28th, Harry M. Donaldson, one of our industrial hygienists, wrote a letter to the Kawecki Berylco people saying that we planned to conduct another survey of the Hazleton plant in the near future and asking what improvements they intended to make in their powder-handling operations, how they were going to make them, and when they would complete them. On August 3rd, I sent a second memo of concern to the regional administrator of the Occupational Safety and Health Administration in Philadelphia—I had sent him one on the same subject on December 22, 1971. I told him again that a potential for serious medical consequences existed in the Hazleton plant, and I enclosed the results of a medical survey of two hundred and nineteen of the plant's employees which had been conducted in November of 1971 by Dr. Homayoun Kazemi, the chief of the pulmonary unit of Massachusetts General Hospital, in Boston. Dr. Kazemi had found symptoms possibly related to beryllium disease in twenty-five cases and significant beryllium disease in half a dozen cases."

I asked Dr. Johnson if he or Donaldson had received any reply from the company, or if any action had been taken with regard to the situation at the plant. He smiled grimly and handed me a memorandum. It was sent to division directors, deputy directors, and assistant division directors; it was headed "Policy Memorandum," was dated September 12, 1972, and was signed by Dr. Edward J. Fairchild II, acting associate director for NIOSH's Cincinnati Operations, who was working under the direction of Dr. Key, the director of NIOSH. It read:

> There have been in the recent past certain communications, verbal as well as written, with distinct overtones of abatement-type language. It is not the intent, nor the policy, of NIOSH to convey to the outside world that our role under the Occupational Safety and Health Act of 1970 is one having authority for enforcement. Rather, we must present an image more in keeping with that of a research agency. The Act distinctly separates enforcement, the primary activity of the Department of Labor (OSHA), from research, the primary activity of HEW (NIOSH).
>
> Accordingly, you are requested to monitor those actions which could evoke connotations of enforcement, especially those which may be so borderline that NIOSH involvement could be appropriate under certain circumstances, yet highly inappropriate under different circumstances. Those

> of your people who may have occasion for involvement should be informed of this policy.
>
> Through authority delegated to me by Dr. Key, and by copy of this memorandum, I assign Dr. [William S.] Lainhart as your contact in the event that questionable issues arise. Thus, if you are in doubt as to whether actions constitute abatement-type language or policy or if activities infringe upon enforcement, you will bring such situations to the attention of Dr. Lainhart. If, then, the issue is not yet resolved, it will be brought to my attention by Dr. Lainhart.
>
> Your cooperation and diplomacy in this delicate, but important, matter of policy is expected and appreciated.

I remembered Dr. Lainhart very well. He had been chief assistant to Dr. Cralley in the old Division of Epidemiology and Special Services, and had been in charge of the medical-environmental teams that surveyed the Tyler plant and other asbestos factories during the middle and late nineteen-sixties. It was Dr. Lainhart who—in March of 1968, shortly before notification of the high asbestos-dust counts taken by his environmental team at the Tyler plant was sent to Dr. Grant and Pittsburgh Corning without any warning that they constituted a high risk of disease and death for the men who were working there—had said to me that the ideal method of tracing the natural history of asbestos disease would be to take a bunch of twenty-year-olds, put them into an asbestos plant where the exact dust level was known, and observe them for the next fifty years or until they died. I had often wondered about that statement, and now, remembering the poor condition of men I had met who had been employed at the Tyler plant for only ten or fifteen years, and the number of deaths from cancer and asbestosis that had occurred among the unfortunate workers at the Paterson factory, I found myself wondering about it again.

Then Dr. Johnson showed me a second memo. It was dated September 13, 1972, and it had been sent to the director of the division. It was headed "Short- and Long-Term 'Epidemiology,'" and was also signed by Dr. Fairchild, who was once again acting under the direction of Dr. Key. It started out:

> Pursuant to our discussion earlier with Dr. Key, as well as followup discussion you and I have had, this will verify for the record the resulting decisions.
>
> Short-term field studies which have been referred to as "firefighting activities" will be assessed prior to obligating resources. We are mutually agreed that you and/or Dr. Johnson will keep Dr. Lainhart apprised of such categorical activities. When necessary I can be brought into the discussions. Dr. Lainhart is aware of this arrangement which is commensurate with his experience and the responsibility of his position.
>
> Also, in connection with our earlier discussions and in keeping with the indication given by Dr. Key to Mr. Edmund Velten, Vice-President, Kawecki Berylco Industries, Inc., there will be developed for NIOSH policy the proposed regulations whereby industry-wide studies are to be conducted. Accordingly, I am requesting that this also be coordinated with Dr. Lainhart so that he is kept informed throughout the process of regulations development.

When I had finished reading the second memorandum, I asked Dr. Johnson to give me an example of a short-term field study, or "firefighting activity."

"Well, in some ways you might call our survey of Pittsburgh Corning's Tyler plant a short-term field study," Dr. Johnson replied.

"And what about the reference to the 'indication' given by Dr. Key to Edmund Velten, of Kawecki Berylco?" I inquired.

"That simply means that the beryllium industry will be consulted by NIOSH with regard to any proposed industry-wide studies involving beryllium," Dr. Johnson said. "The fact is, our criteria document for beryllium was completed and forwarded to the Occupational Safety and Health Administration several months ago. As of this date, however, the Administration has not seen fit to convene an advisory committee on beryllium, much less announce public hearings. Meanwhile, workers in beryllium plants are still being overexposed to beryllium dust and running the needless risk of developing beryllium disease.

Incidentally, Mr. Velten must have complained to Dr. Key about the letter that Donaldson sent to his company about improvements in the Hazleton plant. In any case, Dr. Key wrote Velten a letter containing an apology for the tone of Donaldson's letter."

At this point, I asked Dr. Johnson what was being done about the problem of occupational exposure to beta-naphthylamine and benzidine.

"Beta-naphthylamine, which is commonly known as BNA, and benzidine are aromatic amines—chemicals derived from coal tar—and are used as intermediates in the synthesis of dyes," Dr. Johnson said. "Both are highly potent carcinogens, known to cause bladder cancer, and both are on the list of nine carcinogens that have been rated by the Conference of Hygienists as being too dangerous at any known level of exposure. Neither of them, however, has been so rated by the federal government. Before I go into our involvement with BNA and benzidine, I think you ought to take a look at a study of workers exposed to them in a plant operated by the specialty-chemicals division of the Allied Chemical Corporation, in Buffalo, New York. It was carried out back in 1962, by Dr. Morris Kleinfeld, director of the Division of Industrial Hygiene of the New York State Department of Labor; by Dr. Leonard J. Goldwater, a consulting industrial-hygiene physician; and by Dr. Albert J. Rosso, an associate industrial-hygiene physician in the Division. It was published in the *Archives of Environmental Health* in December of 1965. Dr. Kleinfeld, Dr. Goldwater, and Dr. Rosso studied three hundred and sixty-six workers at the plant in Buffalo, going back to 1912. Half of these men were between twenty and twenty-nine years old when they were first employed at the plant; about a quarter were between thirty and thirty-nine; and the remainder were between forty and fifty. Some of them were exposed to BNA alone; some to BNA and benzidine; some to BNA and alpha-naphthylamine, which is similar to BNA; some to benzidine alone; and some to various combinations of all three compounds."

Dr. Johnson handed me a copy of the study by Dr. Kleinfeld and his associates. From the introduction I learned that cancer of the bladder resulting from exposure to aromatic amines was discovered in Germany in 1895; that between 1905 and 1932 bladder tumors in dye workers were reported in Switzerland, Great Britain, Russia, and Austria; that the first cases of this condition in the United States were reported in 1934; and that the disease had subsequently been recognized in Italy, Japan, and France. Toward the end of the introduction, Dr. Kleinfeld and his associates described a large series of bladder tumors caused by aromatic amines which were reported by Dr. T. S. Scott in England in 1964. Dr. Scott found that of six hundred and sixty-seven persons exposed to the chemicals for more than six months a hundred and twenty-three had developed bladder tumors, and that bladder tumors occurred in fully seventy-one per cent of those who were exposed for thirty years or more. As for the results of the study of three hundred and sixty-six men exposed to aromatic amines at the Allied Chemical plant in Buffalo, Dr. Kleinfeld and his associates reported in the main body of their article that ninety-six of the men had developed bladder tumors—an over-all incidence of slightly more than twenty-six per cent—and forty-six of these had died as a result. "By any reasonable standard it can be stated that the incidence or attack rate for bladder tumors in workers exposed to coal tar dye intermediates is frighteningly high," they concluded.

After I read the study, Dr. Johnson told me that the manufacture and use of BNA had been banned in Switzerland in 1938, and that the production and use of the chemical had virtually stopped in Pennsylvania after 1961, when the state adopted strict regulations for carcinogens. "The British abandoned the manufacture and use of BNA way back in 1952," Dr. Johnson said. "Why our government has not taken similar action is beyond my comprehension, especially in the light of conditions we uncovered recently in Georgia and South Carolina. During the early fifties, the Augusta Chemical Company, of Augusta, Georgia, which had been receiving shipments of BNA from the du Pont people, began manufacturing and using the chemical at its Augusta plant. In 1967, the company was purchased by the Blackman Uhler Chemical Division of the Synalloy Corporation, of Spartanburg, South Carolina, and thereafter BNA was manufactured at the Augusta plant and shipped in drums by truck to Synalloy's Spartanburg plant, where it was used as an intermediate in dye synthesis. Our regional director in Atlanta learned of this activity last fall, and got in touch with officials of the Georgia State Department of Health, who told him they knew of no problem associated with the manufacture of BNA in Georgia. In January of this year, however, we sent three of our people in the Division of Field Studies and Clinical Investigations to take a look at the Augusta plant. They found men shoveling BNA from a large slurry tank into drums, and discovered that none of them had ever been given a urinalysis to determine whether blood or atypical or malignant bladder cells were present in the urine. In April, I visited Spartanburg and Augusta, and told the Synalloy people the medical facts implicating BNA as a potent bladder carcinogen. They claimed that because the chemical was handled in wet slurry form it was not hazardous, whereupon I informed them that the primary route of entry of BNA into the body under these circumstances was not by inhalation but by absorption through intact or unabraded skin. As a result, the production and use of BNA have been discontinued at the Augusta and Spartanburg plants, and Tobias acid—a relatively safe substitute for BNA, which has been available for years, and which costs only a few cents more a pound—is now being used as an intermediate in dye synthesis. We have begun a program of urine cytology on the workers currently employed at the Augusta plant, but since BNA wasn't used there until the late nineteen-forties, and since the latent period for the development of bladder cancer is about twenty years, we probably won't know their true health situation for some time. One big problem we're faced with is that, because of rapid turnover in the work force, past employees exposed to BNA are unavailable for medical followup, so we have only the present workers to evaluate. The thing to remember is that until 1972 no federal occupational-health agency had ever visited either of the plants —and this despite the fact that the hazard of BNA was brought up in congressional testimony in 1970 as justification for passing the Occupational Safety and Health Act, and had been cited repeatedly as being one of the health hazards the government was anxious to control. Moreover, none of the appropriate state occupational-health agencies had ever taken effective action to bring the hazard under control. We're now trying to track down other plants where BNA may be in use, because we've heard that the chemical is being imported into the United States from South America."

Dr. Johnson went on to tell me that the Allied Chemical plant in Buffalo had manufactured BNA until the middle fifties and had received shipments of the chemical from a plant in Pennsylvania from then until 1961, when the Pennsylvania plant was forced by the state to shut down. "Shortly thereafter, the Allied Chemical people discontinued the use of BNA in their Buffalo plant," he said. "However, they continued to handle benzidine, which they not only manufactured and used there but also shipped to the Toms River Chemical Corporation's plant, in Toms River, New Jersey, until sometime in 1971. Now I understand that they are using dichlorobenzidine, which they claim is less hazardous than benzidine. However, we are concerned about dichlorobenzidine because it is known to cause cancer in test animals, and because it is also on the list of carcinogens rated by the Conference of Hygienists as being too dangerous at any level of exposure. In fact, we'll be conducting a field survey of the Buffalo plant next month."

I asked Dr. Johnson what other industrial carcinogens the Division of Field Studies and Clinical Investigations was looking into. He replied that there were many, and that new ones came to his attention almost daily. "Recently, we've been worried about a chemical called bis-chloromethyl ether," he said. "It is formed as an intermediate in the production of anion-exchange resins, which are manufactured by a number of large outfits, including Rohm and Haas and Dow Chemical. Bis-chloromethyl ether has been demonstrated to be a potent lung carcinogen in test animals, and it, too, is on the list of dangerous cancer-producing agents. Last March, at the request of Jack Washkuhn, an industrial hygienist from the San Mateo County Health Department, in California, I went out with Harry Donaldson to visit the Diamond Shamrock Corporation's plant in Redwood City, where the chemical has been encountered as a by-product since 1957. The plant employs only about a hundred men, but we reviewed the cases of four employees who, Washkuhn had learned, had died of lung cancer within the past eight years, and we learned that two former workers at the plant had developed lung cancer but were still alive. The first death among these men occurred in 1964, the second in 1965, the third in 1967, and the fourth in November of 1971. Something that causes us great concern is that the age at death of these four men ranged from thirty-two to forty-eight—which is considered to be under the age at which most lung cancers develop in association with cigarette smoking. Moreover, we are particularly concerned about the worker who died in November of 1971, because he was only thirty-two years old, and he had worked in the plant for only two years—a circumstance indicating that bis-chloromethyl ether may be a far more potent carcinogen than had previously been suspected. Accordingly, we recommended to the company that chest X-rays and sputum cytology be conducted on its workers. A subsequent survey by a special field team from our Salt Lake City facility showed considerable amounts of abnormal cells in the sputum of men who had been working at the plant. As a result, the company has initiated a periodic sputum-cytology screening program of its current and former employees, which is being carried out with the cooperation of Dr. Geno Saccomanno, head of the pathology department at St. Mary's Hospital in Grand Junction, Colorado, who is well known for his sputum-cytology studies of uranium miners."

Dr. Johnson was then called away to attend a meeting. I told him that I hoped to talk with him again soon, and thanked him for his time and help.

"You're entirely welcome," he replied. "By the way, you asked me earlier for an example of the kind of short-term epidemiology, or firefighting activity, we are now required to clear through higher authority. Well, it just occurred to me that our visit to the Diamond Shamrock plant in Redwood City is a perfect one."

Back in New York, I telephoned Dr. Mastromatteo in Toronto, to find out if he had been able to discover anything for me about Holmes Insulation, Ltd., in Point Edward, Ontario. Dr. Mastromatteo told me that not only had he acquired some information about the company but that an industrial hygienist from the Ontario Ministry of Health had visited the company's plant in August to take airborne dust samples and to arrange for X-ray examinations for the workers there. "A few months ago, claims for asbestosis put in by workers employed there began to appear before our Workmen's Compensation Board," he said. "During our August inspection of the plant, we found that, generally speaking, asbestos-dust levels were well over two fibers per cubic centimeter, which is our unofficial standard. Since we always take action if dust levels exceed the five-fiber level, we recommended that the company install better ventilating equipment in order to avoid the use of respirators."

"What did your X-rays show?" I asked.

"They indicated," Dr. Mastromatteo replied, "that out of a work force of between fifty and seventy-five men, six or seven had signs consistent with the development of asbestosis."

During the second week in October of 1972, I called on Dr. Selikoff at Mount Sinai and asked him whether he had heard anything from his British colleagues concerning the marked discrepancy between the findings of Dr. Knox and Dr. Holmes and those of Dr. Lewinsohn in regard to asbestosis in the Turner

Brothers workers. Dr. Selikoff had just returned from a World Health Organization Working Group to Review the Biological Effects of Asbestos, in Lyon, France, and by way of reply he handed me a copy of a paper that Dr. Holmes had delivered at the conference. The paper was entitled "Criteria for Environmental Data and Bases of Threshold Limit Values—Environmental Data in Industry," and the first sentence read, "The need for hygiene standards for airborne asbestos dust based on wider studies than were available to the British Occupational Hygiene Society in 1968 is emphasized." Of the study that he and Dr. Knox had conducted of the workers at the Turner Brothers Rochdale plant, and the data they had furnished the society, Dr. Holmes said, "The information, although the best available at the time, was, to say the least, scanty for the purpose, and some of us who were associated with it have become increasingly concerned with the authority with which it has become invested in the international field." Dr. Holmes ended his paper by noting that in collaboration with Dr. Lewinsohn arrangements were being made to have the information from the factory brought up to date.

Essentially, what Dr. Holmes appeared to have admitted was that the data he and Dr. Knox had furnished the society concerning the incidence of asbestosis in the Turner Brothers workers were questionable. The ramifications of this were, of course, staggering, for if the data were inaccurate, the two-fiber standard for asbestos—which would not even go into effect in the United States until 1976—would be without medical or scientific validity. In the meantime, American asbestos workers had been assured by Secretary of Labor Hodgson that it was safe to breathe air containing five fibers per cubic centimeter, which was higher than the dust concentrations inhaled by a significant number of the Turner Brothers asbestos workers, many of whose chest X-rays had been interpreted by Dr. Lewinsohn as showing abnormal lung changes. All this, of course, could affect the health of tens, if not hundreds, of thousands of asbestos workers, and it had come about simply because two men—a medical doctor and an industrial hygienist—employed by the British asbestos industry had, one way or another, furnished questionable data, which may have misled occupational-health agencies on both sides of the Atlantic. One could only wonder exactly when Dr. Holmes had become "increasingly concerned" about the authority and validity of Dr. Knox's interpretation of the chest X-rays of the Turner Brothers workers. It can hardly have been prior to his appearance in behalf of the asbestos industry at the Department of Labor's public hearings six months earlier, when he declared that a standard of four or five fibers per cubic centimeter would provide adequate protection from disease for asbestos workers in the United States. Was it, then, only after the publication of Dr. Lewinsohn's findings, which suggested that Dr. Holmes and Dr. Knox could have understated by as much as tenfold the extent of asbestos disease among the Turner Brothers workers?

In the middle of October, I went to Washington to talk with Mazzocchi about some of the other health problems that were plaguing American workers. Over the past few years, during which he and his associates in the Legislative Department of the OCAW had been holding conferences for factory workers around the country, they had amassed and published, in a series of pamphlets entitled "Hazards in the Industrial Environment," a great amount of information on the effects of chemicals and physical agents on the health of working men and women. In gathering this information, they had compiled a large dossier of the names, affiliations, and activities of people who were involved in occupational-health matters, and so were known to have "the book" on the medical-industrial complex. I had not seen Mazzocchi since the public hearings in March. He had been much on the go in the interval—flying here and there around the country—and he looked tired.

"You've been studying the Tyler plant, which in many ways is a perfect example of the kinds of problems we face," he told me. "Tyler had all the elements—suppression of occupational-health data, callousness on the part of those in positions of responsibility in industry and government, and danger not only for the workers but also for their families and the community at large—and yet you could be hearing much the same story about any number of factories in this country. You could be studying the Mobil Oil Corporation's refinery in Paulsboro, New Jersey, or the Kawecki Berylco Industries factory in Hazleton, Pennsylvania, or any one of hundreds of other places where the health of workers either has been or is being needlessly endangered. I hardly know where to begin. In fact, I *don't* know where to begin. Steve Wodka once stated the multiplicity of our problems by saying that members of our union don't get black lung, like coal miners, or brown lung, like cotton-textile workers, but rainbow lung, because they're exposed to so many toxic substances."

"Let's start with the Kawecki Berylco plant in Hazleton," I said. "I've already heard something about the situation there."

Mazzocchi asked Wodka to pull out a file on KBI, as the company is known, and told me that the union had represented workers at the factory since the summer of 1969. "By August of 1970, it was apparent that much more information was needed concerning the hazards of beryllium, so the local union representatives initiated plans to hold a conference on the problem in mid-October," he said. "At that conference, the KBI workers told us that beryllium dust was contaminating the Hazleton plant, and that several deaths and current cases of lung disease among them were thought to be related to beryllium exposure." Mazzocchi went on to say that, armed with tapes of the conference, he traveled to Cambridge a few days later and met with Dr. Harriet Hardy, who was assistant medical director in charge of the environmental medical service at MIT, an associate physician at Massachusetts General Hospital, and a renowned expert on beryllium poisoning. Upon hearing the tapes, Dr. Hardy expressed concern about the reports of conditions inside the Hazleton plant and pledged her assistance in analyzing urine samples of the workers there. "Shortly thereafter, an inspector from the Department of Labor conducted a survey of the plant," Mazzocchi said. "The inspection, which the local had requested some sixteen months earlier, was superficial, to put it mildly. For example, no samples were taken to determine the amount of beryllium dust in the air. The inspector did indicate, however, that better exhaust ventilation was required, so that men would not have to wear respirators on routine production jobs. He also criticized the plant for poor housekeeping procedures and for using an open dump for disposing of refuse contaminated with beryllium. The company was given until February 1, 1971, to correct these conditions, but a check with the Department of Labor in March revealed that no followup inspection had been performed to determine whether the company had complied."

Mazzocchi asked Wodka to continue the story of the Hazleton plant, since he had spent considerable time there in the fall of 1970 and the winter of 1971. Wodka—a man in his middle twenties, with brown eyes, curly hair, and a laconic manner—told me that he went to Hazleton in late November of 1970 to gather information on conditions in the KBI plant, and returned on December 8th to attend a union-management health-and-safety meeting. "When I questioned company officials on the unsafe procedures I had heard about, they offered to take me on a tour of the plant," Wodka said. "It proved to be an invaluable experience in assessing unsafe conditions, for almost everything I saw confirmed what the workers had told me. Inadequate ventilation equipment had allowed dust to pile up around machinery and to blow throughout the plant. Another serious problem was the company's excessive reliance on the use of respirators. Instead of installing proper ventilation equipment to reduce beryllium-dust levels, the company had simply designated some of the dusty sections of the plant as respirator areas. As a result, many workers were forced to wear respirators for long periods—a practice that is extremely difficult, if not downright impossible."

A couple of days later, Wodka said, he went to Cambridge to

report what he had learned to Dr. Hardy. "She got on the phone immediately and rounded up a team of beryllium specialists to help us out," he continued. "They included Richard I. Chamberlin, an industrial-hygiene engineer at MIT, who would sample beryllium-dust levels in the plant and study the ventilation system; Dr. John D. Stoeckle, an expert in industrial lung disease at Massachusetts General Hospital, who would conduct a symptom survey of all workers; Dr. Homayoun Kazemi, the chief of the pulmonary unit at Massachusetts General, who would do lung-function tests; and Dr. Alfred L. Weber, a radiologist from the hospital, who would supervise a mass X-ray program and read and interpret the results. When I returned to Washington, I drew up a set of proposals for the next meeting of the union-management health-and-safety committee, on January 5, 1971. The chief provisions were that beryllium-dust levels in the plant be kept at or below the threshold limit value of two micrograms per cubic meter of air; that the union have the right to observe the company's dust-monitoring operations and have access to the results; and that the company finance a thorough industrial-hygiene survey of the plant, to be conducted by the team of specialists designated by Dr. Hardy, which Chamberlin had estimated could be accomplished at a cost of about twenty-one hundred dollars."

Mazzocchi continued the story by telling me that at the January 5th meeting plant officials informed him that they were not prepared to deal with such extensive proposals. "They said they would have to consult higher authority in KBI," he said. "As things turned out, we weren't able to arrange another meeting with the KBI people until March 10th. In the meantime, urine samples were collected from five workers and sent to Dr. Hardy, who had them analyzed at MIT by George W. Boylen, Jr., an expert in industrial-hygiene chemistry. He found varying amounts of beryllium in all the samples, and both he and Dr. Hardy said they believed that beryllium would not be present in the urine of men who were exposed at or below the level of two micrograms per cubic meter of air—the standard recommended by the American Conference of Governmental Industrial Hygienists. As a result, the union local voted unanimously that each of the two hundred and seventy-eight members working in the Hazleton plant support the cost of our health proposals by paying a dollar a month surcharge on membership dues."

Mazzocchi went on to tell me that Wodka conducted an investigation of the State of Pennsylvania's role in enforcing health regulations at the Hazleton plant. "It turned out that the state people had more data on the plant than anyone else," Mazzocchi said. "The quality of the data, however, was another matter. When Steve telephoned Edward Baier, who was then director of the state's Division of Occupational Health and is now the deputy director of NIOSH, he was told that there had been no cases of beryllium disease from the Hazleton plant. Later, we learned that in an annual screening program of KBI workers the Pennsylvania people had been using X-ray equipment that, according to Dr. Hardy, was inadequate for detecting beryllium disease. During his telephone conversation with Baier, Steve learned that the state had taken beryllium-dust counts during its inspections of the Hazleton plant. However, Baier told Steve that he considered the data confidential, and felt that their release to the union would damage the relationship between the state and the company."

Wodka said that he then went to Harrisburg to meet with Baier. "During the meeting, Baier and his staff admitted that they had recently taken readings of beryllium-dust levels in the Hazleton plant, and in some sections they were ten times the recommended standard," Wodka said. "Baier assured me that each man in these contaminated areas had been told to wear a respirator, but he also acknowledged that such protection should be used only for interim control. I asked him why the state had not forced the company to comply with the recommended beryllium standard, and he told me that the KBI people were cooperating. When I again requested access to the state's inspection reports, Baier promised that a copy of the most recent survey, which had been conducted in January, would be sent to the company for transmittal to the union."

Mazzocchi then told me that at the March 10th meeting of the union-management health-and-safety committee, C. Dale Magnuson, the industrial-relations manager for KBI, rejected the union's proposal for a survey of the plant by the team of industrial-hygiene experts that Dr. Hardy had assembled. "According to Magnuson, the company already had a good industrial-hygiene department, and there was no need to bring in another group," Mazzocchi said. "Wodka then asked the KBI people if the State of Pennsylvania had forwarded them a copy of its January inspection for transmittal to the union. They replied that the report had not come in. When Steve produced a letter from Baier stating that the report had been sent to them on March 5th, the KBI people admitted after all that they had received the copy marked for transmittal to the union, but said that they felt under no obligation to turn it over."

The only counterproposal made by the KBI people at the March 10th meeting, Mazzocchi said, was that they would give the union a quarterly report of the company's monitoring data on beryllium-dust levels. "On April 29th, we went back to Hazleton to get the first of these reports, taking along Chamberlin, from MIT, in a consulting capacity. At the meeting, the KBI people were represented by Magnuson; Edmund Velten, one of their vice-presidents; and James Butler, the assistant to the president. The company's report showed only one area in the plant where there were excessive levels of beryllium dust. Upon reviewing the data, however, Chamberlin said that it was difficult for him to interpret the figures without knowing something about the layout of the plant. Velten then suggested a quick tour, to which we readily agreed. At the end of the tour, Chamberlin said he had seen enough to convince him that the company's interpretation of its data on beryllium-dust levels did not give a true indication of the potential hazard in the plant. Then Steve produced copies of the January inspection conducted by the State of Pennsylvania's Division of Occupational Health, and two earlier state inspections of the plant as well, which Baier had finally decided to release the day before. The reports showed serious violations of both state and federal beryllium standards in four specific areas of the factory, going back to 1969. At that point, Velten and the other KBI officials agreed to our key proposal that a union-designated investigation team be allowed to conduct industrial-hygiene and clinical surveys at the plant, and that the costs be shared equally by KBI and the union. In May, we got the Pennsylvania people to agree to take X-rays using equipment adequate for the detection of beryllium disease of all the workers in the Hazleton plant, and to allow Dr. Weber to assist in the interpretation of the X-rays. In mid-September, an inspection of the Hazleton plant conducted by the Occupational Safety and Health Administration still showed excessive levels of beryllium dust. The Administration fined KBI six hundred dollars and gave it a month to clean things up. Then, in November of 1971, Dr. Kazemi, who is one of the world's leading experts on the detection and diagnosis of beryllium disease, conducted a medical survey of the Hazleton workers. Dr. Kazemi found that, out of two hundred and nineteen workers, twenty-five had symptoms possibly related to beryllium disease and seven of these men had lung abnormalities of such magnitude that they shouldn't have been working. Since then, four of the seven have been definitely diagnosed as having beryllium disease. In addition, two other men from the plant, who had been examined at Massachusetts General, were diagnosed as suffering from the disease."

"What happened as a result of the fine and the date for cleaning things up?" I asked.

Mazzocchi gave a short, harsh laugh. "As of this date, more than a year later, the Administration people haven't seen fit to make a followup inspection of the Hazelton plant," he told me. "No doubt they assume the KBI people have complied with their order. Sounds like Tyler all over again, doesn't it?"

After a short coffee break, Mazzocchi said, "Now let me tell you about the Mobil refinery in Paulsboro, which is just southwest of Camden. It employs about a thousand workers, and it manufactures a whole spectrum of oil products, including heating oil, lubrication oil, gasoline, and aviation fuel. For a long time, members of our Local 8-831 had been complaining about health and safety conditions there, and on October 11, 1971, we filed a complaint with Alfred Barden, acting regional administrator of the Occupational Safety and Health Administration in New York City, requesting an imminent-danger inspection of the facility. The petition presented Barden with a partial list of health hazards at the refinery, including exposure to asbestos, sulphuric-acid fumes, phenol fumes, carbon-monoxide gas, tetraethyl lead, caustic soda, benzene, cumene, carbon tetrachloride, and chlorine. Since the Administration's regulations permit employees to request that a third party accompany its own representatives on walk-around inspections, the local asked for Steve Wodka, who had assisted them in filing the complaint. The first imminent-danger inspection began on October 15th, resumed on the 19th, and lasted through October 22nd, and Steve's effect on it was soon evident. Even the inspectors remarked that he provided a valuable extra set of eyes during the walk-around. On the other hand, the Mobil people grew increasingly irritated by his presence—so much so, in fact, that on October 22nd the plant manager told Steve that he hoped to have him removed from the premises."

Wodka then described some of the conditions that were discovered during the first part of the inspection of the refinery. "We found workers installing asbestos insulation, similar to the product manufactured in the Tyler plant, on boilers and pipes in various areas of the plant," he said. "In the asbestos shop, where the insulation was cut to size, the men were wearing surgical-type paper masks provided by the company, which are virtually useless for protection against toxic dusts. Moreover, there was absolutely no ventilation equipment in the cutting shop. Elsewhere in the refinery, we found places where men were being exposed to toluene [an aromatic compound similar to benzene], to excessive noise, to welding fumes, and to hydrogen-sulphide gas—a highly toxic substance that is given off in the sulphur plant, where sulphur is removed from the crude oil."

Wodka went on to tell me that when the inspection resumed, on November 8th, he was denied entry to the refinery. "I was told by John Kearney, the assistant regional administrator, that a decision had been made to bar me from accompanying the walk-around any farther," Wodka said. "As a result, I left the plant, under protest. The inspections continued intermittently for several more weeks, but their quality deteriorated, for by giving management advance notice of the areas they wished to tour, the inspectors also gave management an opportunity to reduce operations that were generating harmful fumes and dust. In this way, the Mobil people were able to dominate the inspection process. As a result, the imminent-danger inspection of the Paulsboro refinery was an almost total failure. Take the problem of exposure to asbestos. The Administration failed to issue a citation for excessive asbestos dust, yet when Dr. Selikoff examined and X-rayed nineteen workers who were employed at the refinery as welders, pipe-coverers, boilermakers, and brick-layers, he found that more than half of them showed X-ray abnormalities consistent with asbestosis. When the Administration, on January 28, 1972, issued citations to Mobil, they carried with them fines totaling seventy-three hundred and fifty dollars for three hundred and fifty-four safety-and-health violations, or about twenty dollars per violation. Only twelve of the violations involved occupational-health standards, one of which was for an unsanitary water cooler."

Mazzocchi said that similar performances by the Occupational Safety and Health Administration had occurred in recent months at a chemical plant in Alabama and at an oil refinery in Kansas. "These are just a few of the many instances in which the Administration has failed to enforce the provisions of the Occupational Safety and Health Act," he said. "By the way, did I tell you that when the Administration people fined KBI six hundred dollars for excessive beryllium dust in the Hazleton plant they also fined the company the grand total of *six dollars* for allowing food to be eaten and stored where toxic materials were present? That was just a few weeks after they fined Pittsburgh Corning the sum of two hundred and ten dollars for so-called 'nonserious violations' at the Tyler asbestos plant. Such fines are ridiculous, of course, but the Administration people don't stop there. They rub salt into the wound by being secretive with what, under law, is public information. We face delays and denials practically every time we ask them for copies of citations, notice of proposed penalties, and inspectors' reports of plant surveys. And, remember, we're trying to exercise the rights of our hundred and eighty thousand-odd members through an extensive union bureaucracy. Imagine what the chances are for unorganized workers, who account for seventy-five per cent of the total labor force in this country!"

At this point, I asked Mazzocchi if he had ever heard of Dr. Mitchell Zavon, Assistant Health Commissioner for Cincinnati.

"Yes, I've heard of him," Mazzocchi replied. "Why do you ask?"

"Because I understand he has recently voiced opposition to some proposed city regulations that would ban the spraying of asbestos insulation in building construction."

"We have a whole file on Dr. Zavon," Mazzocchi said. "I'll have it Xeroxed for you. In 1969, the House Committee on Government Operations wrote a report on Dr. Zavon's activities with regard to the No-Pest strip, manufactured by the Shell Chemical Company, a division of the Shell Oil Company. According to the report, Dr. Zavon had been a consultant to Shell Chemical during a six-year period—1963 to 1969—when he was also a consultant to the Department of Agriculture's Pesticides Regulation Division, which had registered the insecticide strip for use in restaurants and homes. During that time, Dr. Zavon conducted tests for the company which showed the strips to be safe. When it was demonstrated later that the strips could leave unsafe chemical residues on exposed food, the

Department of Agriculture required that they bear a warning label. In the end, the Department informed the Committee on Government Operations that it was referring questions of possible conflict of interest involving Dr. Zavon to the Department of Justice. Nothing came of it, however."

When I left Mazzocchi's office, I was carrying a manila envelope that contained the file on Dr. Zavon and copies of letters, petitions, inspections, and surveys relating to KBI's Hazleton plant and Mobil's Paulsboro refinery. On the plane back to New York, I read the report on Dr. Zavon, which substantiated what Mazzocchi had told me. (Later, I learned that in March of 1973 NIOSH awarded a contract for $205,873 to the Agatha Corporation—a private medical consulting firm headed by Dr. Zavon—for developing health criteria on carbon tetrachloride, chloroform, ethylene dichloride, methylene chloride, tetrachloroethylene, and 1, 1, 1-trichloroethane, which can be synthesized from precursor chemicals obtained from petroleum.) I also glanced through the annual stockholders' report put out by KBI in 1971. It was full of handsome photographs and interesting statistics. It listed principal plants in Hazleton, Boyertown, and Reading, Pennsylvania, and in Wenatchee, Washington. It listed company subdivisions and subsidiaries in Boston; Springfield, Oregon; Revere, Pennsylvania; Thomaston, Connecticut; Yonkers, New York; and Palisades Park, New Jersey. And it listed international operations in France, Germany, the Netherlands, and Great Britain. In a letter to the shareholders, dated March 2, 1972, and signed by Joseph C. Abeles, chairman of the board, and by Walter R. Lowry, president, there was this paragraph:

> No industrial enterprise today can afford to ignore the growing insistence upon clean air, clean water, and safe, healthful places of work as manifested in antipollution and job safety and health legislation. Considerations of environmental quality which have prompted us to make substantial expenditures on equipment and staff to control pollution

> and promote occupational health have, in our view, assumed a permanent place in the conduct of a business such as KBI's. We expect to continue to invest in environmental quality to preserve the gains we have made and take advantage of improvements in control technology as they become available.

A few days later, in the October 29, 1972, edition of the *Times*, I came across a story written by Homer Bigart which appeared under the headline "Lung-Disease Problem, Traced to Beryllium Refinery, Plagues Hazleton, Pa." Bigart's article began:

> In the grim year of 1956, when unemployment in this worn-out coal town hovered near 20 per cent and the region was one of the most depressed in Appalachia, there was general rejoicing when the Chamber of Commerce enticed a beryllium refining plant to settle four miles east of here.
>
> There was only one discordant voice. Dr. Herman H. Feissner, Jr., who has a tiny walk-up office over a store in nearby Freeland, began cautioning his patients that it might be dangerous to work at the plant. Nobody paid much attention.

After saying that nine present or former employees of the plant were suffering from chronic berylliosis, which he described as "a rare disease that involves a slow but progressive—and apparently irreversible—deterioration of the lungs," Bigart returned to the subject of Dr. Feissner:

> The tall, white-haired physician was graduated from Lehigh University in 1928 and from Jefferson Medical College, Philadelphia, in 1932. Now in his late sixties, he spends much of his time working with retarded children at the White Haven State School and Hospital.
>
> However, he still practices medicine and among his recent patients were several beryllium refinery employees. "At least five," he said, had symptoms of berylliosis.
>
> Dr. Feissner's first encounter with the beryllium industry came soon after Kawecki Berylco acquired an old Lehigh Valley Railroad machine shop and roundhouse a few miles east of Hazleton and began converting beryl ore into

> beryllium. Kawecki Berylco is a major company in the specialty metals field, with sales totaling $70 million last year.
>
> What disturbed Dr. Feissner was the knowledge that the corporation's plant in Reading, Pa., had been cited in several lawsuits charging negligence in exposing workers and nearby residents to unsafe levels of toxic dust.
>
> Dr. Feissner said he was reproached by a local radiologist, Dr. Edgar L. Dessen, for "telling my patients a little too much about beryllium poisoning." Dr. Dessen was the leader of a Chamber of Commerce drive to bring new industry to Hazleton, a campaign so successful that unemployment is now down to 4.5 per cent.
>
> Dr. Dessen confirmed that he had spoken to Dr. Feissner.
>
> "He (Dr. Feissner) was telling people beryllium was a toxic material at a time the company was hiring men," Dr. Dessen said. "I felt it was unfair to people who wanted work. The plant was designed under Atomic Energy Commission specifications to keep irritants out of the air. With stringent precautions, the men would be properly protected."
>
> Dr. Dessen said he had been a director of Kawecki Berylco but did not stand for re-election this year. He said his time was taken up with other matters, including the chairmanship of the American College of Radiology's task force on pneumoconiosis, a generic term for lung diseases caused by dust.
>
> "I'm automatically suspect for having been a director," Dr. Dessen said, "but I consider myself a physician first."

When I later made inquiries about Dr. Dessen, I learned that, in addition to being a former director of KBI, he had been paid by the company for many years to read and take X-rays of employees at the Hazleton plant.

By this time, I was beginning to understand how multiple and intricate were the reasons for the appalling casualty rate in the nation's workplaces, and how intertwined and pervasive were the activities of the medical-industrial complex, which was apparently bent on perpetuating the situation. What seemed more and more incredible to me as the months passed, however, was how such a situation could be—indeed, was being—tolerated at the highest levels of the federal government. Then, in the second week of November, Sheldon Samuels sent me a copy of a speech he had delivered before a joint session of the American Society of Safety Engineers and the National Safety Conference, in Chicago, on November 1st. After referring to Secretary Richardson's estimate of a hundred thousand deaths annually from occupational disease, Samuels addressed himself to the very question that had been troubling me.

"The economics of the situation are very simple," Samuels said. "Nearly half of the male blue-collar work force is afflicted with chronic—and no doubt partly work-related—diseases that are largely paid for by the worker and the community as a whole. Even if all of the identifiable costs were placed on the employer, we cannot always be sure that it would not be cheaper for the employer to replace dead workers than to keep them alive. It may even be profitable, if only dollars and cents are counted. In the case of chronic occupational disease, it may be cheaper for any nation to sacrifice a life that has already achieved peak productivity." Samuels went on to say that in recent months he had learned of at least four plants in which beta-naphthylamine and benzidine were used without proper controls by employers well aware of the probable death from cancer of a third or more of the workers exposed to them. "Because its priorities are determined on a crude cost-benefit basis, however, the federal government has refused to adopt standards for these and seven other carcinogens," he declared.

Later in his speech, Samuels reminded his listeners that most of them worked for companies that were members of the United States Chamber of Commerce, the National Association of Manufacturers, and similar groups. "Read the record of the recent Senate and House oversight hearings of the committees on small business and labor," he said. "The Chamber says that workers face no greater risk than slipping in a bathtub. I only wish the situation were confined to the greedy bluster of such moral midgets. More serious is the subversion of government for profit. The National Association of Manufacturers has a full-time representative—office, phones, even a government expense account—at the very heart of the Occupational Safety and Health Administration operations. Political industry appointments are now made at the lowest levels. Industry standard-setting organizations, such as the American Society for Testing Materials, are attempting to bypass the standard-setting mechanisms in the Act. Industrial consultants dominate the contract route the federal government has taken in lieu of adequate staff in its standard-setting operations."

Of particular interest to me was Samuels' claim that industrial consultants were receiving contracts that in effect allowed them to usurp the standard-setting provisions of the 1970 Act—which had clearly intended that standards be developed and recommended by NIOSH and then promulgated and enforced by the Occupational Safety and Health Administration. I was already familiar with the activities of Arthur D. Little, Inc., in helping to set the questionable compromise standard for asbestos, and a few days earlier I had received information concerning another industrial-consulting firm under contract to the government, which had attempted to play a similar role. This had occurred when Mazzocchi called to suggest that I telephone Dr. Jeanne M. Stellman, who is assistant for occupational health to the president of the Oil, Chemical, and Atomic Workers, at the union's headquarters, in Denver. "Ask her to tell you the story about benzene," Mazzocchi said.

When I called Dr. Stellman, she explained, to begin with, that benzene—a colorless liquid—is produced as an integral part of refining oil and gasoline. "It is probably handled by more than a third of the hundred and eighty thousand-odd members of the Oil, Chemical, and Atomic Workers," she said. "And because benzene is also widely employed in the rubber, cement, and plastics industries, additional thousands of workers come into contact with it there. The trouble with benzene, as has long been known, is that it is a noxious poison, whose fumes, when inhaled, can induce blood changes—anemia and leukemia.

Indeed, benzene is considered so dangerous that both the American National Standards Institute and the International Labour Organisation have recommended that in any given volume of air, liquid, or solid material it should be present only in the ratio of ten parts per million—which comes to one-thousandth of one per cent—with a ceiling of twenty-five parts per million. The American Conference of Governmental Industrial Hygienists, however, has recommended a standard whereby benzene can be present in a ratio of twenty-five parts per million, with a ceiling of fifty parts per million. I tell you all this as background to the fact that in order to develop criteria for the establishment of an official federal standard for benzene a twenty-three-thousand-dollar contract from NIOSH was awarded last May to George D. Clayton & Associates—an industrial-consulting firm in Southfield, Michigan."

On October 10th, Dr. Stellman continued, Clayton & Associates called an informal meeting of a benzene committee it had established. The meeting was held at the William Penn Hotel in Pittsburgh, where the Industrial Health Foundation, Inc., was then holding its annual conference. "As a member of the committee, I was invited to attend," she said. "Clayton & Associates called the meeting at that time and place simply because several of the committee's members were attending the foundation's annual conference. The idea was to have us give our opinion of a preliminary draft of the document on benzene criteria they had prepared for NIOSH. In addition to Clayton, Robert G. Keenan, vice-president and director of laboratories for Clayton & Associates, who was formerly head of the analytical laboratories at the Bureau of Occupational Safety and Health, and myself, the members of the committee included Howard L. Kusnetz, who was formerly director of the Bureau's Division of Occupational Injury and Disease Control and is now with the industrial-hygiene department of the Shell Oil Company; Dr. Horace W. Gerarde, who was formerly chief toxicologist at the Esso Research and Engineering Company and is now at Fairleigh Dickinson University, in New Jersey; a representative of the Manufacturing Chemists Association and one from the American Steel Institute; and Louis Beliczky, the director of industrial hygiene for the United Rubber, Cork, Linoleum, and Plastic Workers of America, who could not attend. When I read the preliminary draft of the benzene document prepared by Clayton & Associates, I was shocked to see that it was proposing a benzene standard identical to the one recommended by the Conference of Hygienists, which is two and a half times as high as the standard for benzene recommended by the American National Standards Institute and the International Labour Organisation. I told the committee that there was no way labor could live with such a standard. After a whole day of arguing, I finally walked out of the meeting, took a plane to Washington, and told Sheldon Samuels what was going on. Samuels promptly lodged a complaint with Dr. Marcus M. Key, the director of NIOSH, and, as a result, Clayton & Associates revised its preliminary document on benzene criteria, and has since recommended the more stringent standard for benzene of ten parts per million. The point is that the firm obviously called the Pittsburgh meeting in the hope of obtaining approval for a compromise standard for benzene which, while it might not adequately protect workers against anemia and leukemia, would not ruffle anyone's feathers."

Since Dr. Stellman's account of how Clayton & Associates had handled the NIOSH contract for developing criteria on benzene was closely followed by Sheldon Samuels' speech, I started to look into this aspect of the medical-industrial complex by examining a book entitled *NIOSH Contract and Research Agreements*, which the Department of Health, Education, and Welfare had published in September of 1972. Without much difficulty, I determined that out of a hundred and forty-eight contracts for research on occupational safety and health which NIOSH had either let or renewed in the fiscal year 1972, eight had been awarded to Clayton & Associates. In addition to the contract for benzene, the firm had received contracts for developing documents on criteria for toluene, chromic acid, and trichloroethylene; for improving methodology for monitoring heat-stress standards; for writing a manual of good industrial-hygiene practices for metal welding and cutting; for setting up a basic course on particulate contaminants for presentation at the college level; and for publishing the third edition of a Public Health Service manual entitled "The Industrial Environment—Its Evaluation and Control."

As I read through the listings, I noticed that the project officer for the contracts awarded to Clayton & Associates for the development of criteria for benzene, toluene, chromic acid, and trichloroethylene was Dr. Charles H. Powell, who is NIOSH's assistant director for Research and Standards Development, at the Institute's headquarters, in Rockville, Maryland. Considering Dr. Powell's position, this in itself did not appear to constitute anything out of the ordinary. Nor was it particularly surprising that Dr. Powell was listed as project officer for five out of seven other contracts let by NIOSH, for the development of criteria documents on mercury, sulphur dioxide and sulphuric acid, toluene diisocyanate, fibrous glass, cadmium, arsenic, and parathion. What did seem somewhat unusual, however, was the fact that four of the five contracts for which Dr. Powell was project officer had been awarded to the consulting firm of Tabershaw-Cooper Associates, Inc., of Berkeley, California. (This firm had also received a fifth contract, for which Dr. Powell was not listed as project officer, for developing criteria on arsenic.) The first contract for which Dr. Powell was project officer was awarded to the Sequoia Group, also of Berkeley, for developing a parathion-criteria document. The contract for a criteria document on cadmium was awarded to the University of Cincinnati's College of Medicine. Thus, out of the eleven most crucially important contracts let by NIOSH in the fiscal year 1972—those for the development of criteria designed to enable the federal government to set specific standards for hazardous substances—Dr. Powell was listed as project officer for four contracts, totaling $115,328, that had been awarded to Clayton

& Associates, and for four contracts, totaling $156,885, awarded to Tabershaw-Cooper Associates. In view of the way Clayton & Associates had handled the benzene-criteria document in Pittsburgh, I found this a bit disconcerting. And in view of the fact that Dr. W. Clark Cooper, a partner in Tabershaw-Cooper Associates, had published the results of two studies on exposure to fibrous glass in separate articles in the *American Industrial Hygiene Association Journal*, and that both studies had been supported by the National Insulation Manufacturers Association, of which PPG Industries, the Owens-Corning Corporation, and Johns-Manville (all large producers of fibrous glass) either are or were members, I found it downright eerie, especially when I recalled that the trail of the medical-industrial complex, which I had been following for nine months, started at the Tyler asbestos plant, which was owned by the Pittsburgh Corning Corporation, a joint venture of PPG Industries and the Corning Glass Works, which, in turn, had established and still partly owns Owens-Corning.

But if the labyrinth in which the medical-industrial complex operated now seemed to branch into a new network of tunnels, a truly remarkable example of the elements making up this complex appeared in black and white several days later, when I received in the mail a copy of a book entitled *Industrial Environmental Health: The Worker and the Community*, newly published by the Academic Press, of New York and London. The book, whose frontispiece proclaimed that it was sponsored by the Industrial Health Foundation, Inc., identified its editor as "Lester V. Cralley, Environmental Health Services, Aluminum Company of America, Pittsburgh, Pennsylvania," and listed its associate editors as "Lewis J. Cralley, National Institute for Occupational Safety and Health, Cincinnati, Ohio; George D. Clayton, George D. Clayton & Associates, Southfield, Michigan; and John A. Jurgiel, Industrial Health Foundation, Inc., Pittsburgh, Pennsylvania." The real and apparent tieups in this editorial roster were startling, to say the least, and they went so many different ways at once as to be almost too much to comprehend. However, after making a few telephone calls and toying with various possibilities, I came up with a way of looking at them which amounts to a riddle.

The riddle begins with the fact that Lester V. Cralley, of the Aluminum Company of America, is a twin brother of Dr. Lewis J. Cralley, who in 1970, while he was director of the Bureau of Occupational Safety and Health's Division of Epidemiology and Special Services, rejected a proposed study of the biological effects of coal-tar-pitch volatiles on workers in the aluminum industry (coal-tar-pitch volatiles being cancer-producing agents to which thousands of coke-oven workers are exposed), on the ground that it was not presented as an industry-wide study, an action that was reversed by the new people in NIOSH, who approved a study of the effects of coal-tar-pitch volatiles on aluminum workers for the fiscal year 1973, which study was temporarily shelved at NIOSH headquarters when some aluminum-industry officials voiced objections to not having been consulted during the planning stages of the investigation, an event that appeared to repeat the previous action taken by Lewis J. Cralley, who, now retired from government service, participated in writing up the health effects of benzene as a paid consultant to Clayton & Associates, which is the firm that received a hundred-and-sixty-thousand-dollar contract from NIOSH to produce the third edition of "The Industrial Environment—Its Evaluation and Control," not to be confused with *Industrial Environmental Health: The Worker and the Community*," which was sponsored by the Industrial Health Foundation, Inc., of Pittsburgh, where George D. Clayton and Robert G. Keenan, of Clayton & Associates, called a meeting of the firm's committee on benzene in order to review a criteria document proposing an inappropriate and later discredited compromise standard for benzene, at the Hotel William Penn, on October 10, 1972, which was the very time and place of the annual meeting being held by the Industrial Health Foundation, Inc., which employs John A. Jurgiel, an associate editor of *Industrial Environmental Health: The Worker and the Commu-*

nity, and also employs Dr. Paul Gross, who, in addition to having testified for Johns-Manville in a number of workmen's-compensation cases, is the director of research laboratories of the Industrial Health Foundation, Inc., which is a new name for the old Industrial Hygiene Foundation of America, Inc., the self-styled "association of industries for the advancement of healthful working conditions" that, entirely financed by industry, including Johns-Manville, was retained by Pittsburgh Corning in the summer of 1963 to evaluate the asbestos-dust hazard at its newly acquired plant, in Tyler, Texas, where, during the next eight years, several more evaluations of the hazard were made, including two by the Bureau of Occupational Health's Division of Epidemiology and Special Services, which showed airborne asbestos-dust levels at the Tyler plant to be grossly out of control, a fact that was not only not made known to the men who worked in the plant—many of whom had inhaled asbestos dust for years without even respirator protection—but never evaluated in terms of the incredible disease-and-death hazard it posed for these men by anyone in the Division of Epidemiology and Special Services, including its director during this period, Dr. Lewis J. Cralley, who, as associate editor of *Industrial Environmental Health: The Worker and the Community*, wrote a section of the book entitled "Epidemiologic Studies of Occupational Disease," containing a five-page chapter on asbestos that described in detail some studies of disease among asbestos workers conducted by Dr. John Knox and Dr. Stephen Holmes, of the Turner Brothers Asbestos Company, who later furnished the British Occupational Hygiene Society with data that appear to have underestimated by as much as tenfold the incidence of asbestosis among the Turner Brothers workers, and by Dr. John Corbett McDonald, of McGill University, in Montreal, whose research on mortality among asbestos miners and millers was financially supported by the Quebec Asbestos Mining Association, but that failed to mention either the study showing the disastrous mortality experience of the asbestos-insulation workers conducted by Dr. Selikoff and Dr. Hammond or Dr.

Cralley's own unaccountably uncompleted study showing an appalling rate of death from asbestos disease among asbestos-textile workers, all of which, in turn, was accepted for publication by his brother, Lester V. Cralley, assistant director of Environmental Health Services of the Aluminum Company of America, who is the editor of *Industrial Environmental Health: The Worker and the Community*, which is the book that inspired the riddle.

Shortly before Thanksgiving of 1972, I telephoned Dr. Johnson, at NIOSH's Division of Field Studies and Clinical Investigations, in Cincinnati, and asked him to tell me about a survey that he and some associates from the division had conducted in October at the Allied Chemical Corporation's plant in Buffalo. "We believe that there are still some problems with the company's benzidine operation, and that there are major problems with its dichlorobenzidine operation, which is a relatively open system," Dr. Johnson said. "The Allied Chemical people have chosen to treat dichlorobenzidine, which has been proved to be carcinogenic in test animals, as a toxic substance—not as a potential human carcinogen, as has been recommended by the English, who say there is no known safe level of exposure to dichlorobenzidine, and the American Conference of Governmental Industrial Hygienists, who say that there should be an absolute minimal exposure to that chemical. For this reason, we're particularly concerned about employees at the Buffalo plant, who have had past exposure to beta-naphthylamine and benzidine, and who are currently being exposed to dichlorobenzidine. Incidentally, I have just come back from North Haven, Connecticut, where I talked with industrial hygienists from the Upjohn Company. In 1962, the Upjohn people bought a plant in North Haven that was owned by the Carwin Company, which had been a manufacturer of benzidine since the late nineteen-forties. The Upjohn people discontinued the benzidine operation in 1963, and their hygienists told us the other day that there had been no problems among the workers. However, we then made a check at the Connecticut State Tumor Registry, in Hartford, and discovered that there had already been six deaths from bladder cancer among workers employed at the old Carwin plant. For this reason, we are concerned by the fact that one hundred and seventy-one employees of the plant, whose onset of exposure to benzidine, dichlorobenzidine, and other aromatic amines occurred more than sixteen years ago, are no longer employed there, and are consequently not included in any existing program of medical surveillance."

Dr. Johnson went on to tell me that on November 9th he had addressed a meeting of the National Tuberculosis and Respiratory Disease Association, in Houston, where he presented some new information concerning the men who worked at Pittsburgh Corning's asbestos plant in Tyler. "Using data from Dr. Selikoff's and Dr. Hammond's mortality study of the nine hundred and thirty-three men who worked at the Union Asbestos & Rubber Company's Paterson, New Jersey, factory between 1941 and 1945, where exposures were similar to those incurred at Tyler, we now estimate that there will be between one hundred and two hundred excess deaths from asbestos-related cancer among the eight hundred and ninety-five men who worked at the Tyler plant," Dr. Johnson said. ("Excess deaths" are deaths beyond the number that the standard mortality tables would project.) "Late last summer, I sent memorandums to various regional administrators of the Occupational Safety and Health Administration, giving the names and addresses of other asbestos plants where I had reason to suspect that, because of data we had found buried in the files, there were problems of overexposure that should be investigated. I have since been informed by one of the assistant directors at NIOSH headquarters, however, that such memos could constitute an embarrassment to our director, Dr. Key, who is apparently anxious to maintain an image of NIOSH as a pure-research agency. In effect, I have been cautioned against alerting the government's enforcement agency to situations where there might be a disease-and-death hazard. I now intend to leave NIOSH at the end of June, when my tour of duty with the government is over, because I've come to the conclusion that I'll be better able to function as a medical doctor in some other atmosphere."

I called Dr. Selikoff in December and asked him to comment on Dr. Johnson's estimate of future mortality among the Tyler workers, and he told me that he thought the estimate would turn out to be too low. "It must be remembered that only thirty-one years have passed since the onset of exposure among the men who went to work at the Paterson factory," Dr. Selikoff said. "It must also be remembered that the excess risk of lung cancer in men exposed to asbestos increases year by year. For example, a man at twenty-five years from onset of exposure has an increased risk over a man at ten years from onset. This risk is greater at thirty years, and even greater at forty. On the basis of the mortality study of the Paterson workers, I expect a dismal future for many of the men who worked in the Tyler plant. In fact, I anticipate that there will be a hundred and fifty excess deaths among them from lung cancer, fifty excess deaths from mesothelioma, forty-five excess deaths from cancers of the colon, rectum, stomach, and esophagus, and fifty excess deaths from asbestosis. In other words, almost three hundred, or roughly a third, of these men will probably die unnecessarily early deaths."

PART FIVE

A Question of the Patient's Rights

During the early part of 1973, occupational-health problems besides those associated with asbestos began to receive increasing public attention. On January 3rd, the Buffalo *Courier-Express* carried an article stating that chemicals used at the Allied Chemical plant in Buffalo had been implicated in a dozen recent cases of cancer, some of them fatal, among workers in the factory. On December 29th, a petition had been filed with the Occupational Safety and Health Administration by the Nader Health Research Group and the Oil, Chemical, and Atomic Workers International Union requesting that a zero level of exposure for ten carcinogens be set through a temporary emergency standard to be issued under the authority of the Occupational Safety and

Health Act. A press release issued on the same day read, in part:

> Approximately 100,000 American workers die each year as a result of occupational diseases. As more is learned about the origins of cancer, it becomes clear that thousands of worker deaths are caused by exposure to carcinogenic chemicals in the workplace.
>
> The Health Research Group and the Oil, Chemical and Atomic Workers Union are thus petitioning the Department of Labor to promulgate emergency temporary standards to eliminate human exposure to the following 10 cancer-causing chemicals in order to protect the lives and health of American workers:
>
> 2-Acetylaminofluorene
> 4-Aminodiphenyl
> Benzidine and Its Salts
> Bis-Chloromethyl Ether
> Dichlorobenzidine and Its Salts
> 4-Dimethylaminoazobenzene
> Beta-Naphthylamine
> 4-Nitrodiphenyl
> N-Nitrosodimethylamine
> Beta-Propiolactone

When I telephoned Dr. William Johnson at NIOSH toward the end of the month and inquired about these developments, he informed me that on January 24th he had called Dr. Albert J. Rosso, who is associate industrial-hygiene physician at the Buffalo office of the New York State Department of Labor's Division of Industrial Hygiene, and that Dr. Rosso had revealed that no followup inspection of the Allied Chemical plant had been conducted by the state since the NIOSH survey in October of 1972, which showed that men working in the factory were at risk because of exposure to benzidine and dichlorobenzidine. On February 9th, the Assistant Secretary of Labor published a notice in the *Federal Register* acknowledging receipt of the petition on the ten carcinogens, and requesting additional information from interested parties. In response to the request, fifty written comments were received during the next few months. Some of them were rather interesting. Dr. Harold Golz, the director of medical-environmental affairs for the American Petroleum Institute, said that the petition should be denied, on the ground that the proposed rules were "unrealistic, technically unfeasible, and inconsistent" and that "zero tolerance is a philosophical concept and an objective that is neither achievable nor necessary." The Benzidine Task Force of the Synthetic Organic Chemical Manufacturers Association asserted that workers were being adequately protected, and then, curiously, went on to say that if tumors did occur they were removed "long before malignancy is expected to develop." Sam MacCutcheon, corporate director for safety and loss prevention of the Dow Chemical Company, said that the suggested standard would serve as a harassment and would dilute present cooperative efforts between industry and government agencies. MacCutcheon went on to say that studies related to bis-chloromethyl ether were in progress, and that, because of their importance and the impact they would have, the chemical should be removed from consideration until they were completed. Bis-chloromethyl ether was also much on the minds of people at the Rohm and Haas Company, in Philadelphia. Frederick C. Moesel, Jr., assistant secretary of the firm, said that exhaustive epidemiological studies on the chemical were under way, and that his company expected results showing a no-effect level well above one part per billion.

On May 3rd, having assessed the petition and the fifty written comments, the Administration issued an emergency temporary standard consisting of strict work practices regarding the manufacture and use of fourteen carcinogens, including bis-chloromethyl ether and the nine others that had been listed in the petition. According to the Administration, workers were being exposed to the fourteen chemicals, such exposure posed a grave danger to them, and the emergency standard was necessary to protect their health until permanent standards could be promulgated six months hence, as the law required.

The subject of bis-chloromethyl ether came to my attention again in the middle of June, when Sheldon Samuels sent me a copy of an article entitled "Lung Cancer in Chloromethyl Methyl Ether Workers," which had been published in the May 24, 1973 issue of the *New England Journal of Medicine* by three

Philadelphia physicians—Dr. W. G. Figueroa, of the Germantown Dispensary and Hospital's pulmonary-disease section; Dr. Robert Raszkowski, of Temple University's School of Medicine; and Dr. William Weiss, of the Department of Medicine of Hahnemann Medical College. (Bis-chloromethyl ether is a contaminant by-product occurring with chemical reactions that take place in the production of chloromethyl methyl ether.) The article began by saying that in 1962 the management of a chemical-manufacturing plant employing about two thousand workers "became aware that an excessive number of workers suspected of having lung cancer were being reported in one area of the plant, and turned to a chest consultant, who recommended a program to establish the degree of risk by semiannual screening." This screening program, which included chest X-rays, was in progress for the next five years, the article said, and during that time the plant management "made a careful investigation of the work histories in several men whose lung cancers developed while they were working in the area under suspicion, and concluded that the only common denominator was exposure to chloromethyl methyl ether (CMME)."

What action, if any, the plant managers took with regard to the conclusion they had reached by 1967 remains a mystery, for the article continued:

> Management is as yet unable to provide exact information on the exposure of the employees to CMME. Further interest that CMME could be a carcinogen was stimulated by . . . a 44-year-old man admitted to Germantown Dispensary and Hospital in December, 1971, because of cough and hemoptysis. A detailed occupational history revealed that he was a chemical operator who had been exposed to CMME for 12 years. The patient stated that 13 of his fellow workers had lung cancer, and he suspected that this was his diagnosis. All had worked as chemical operators in the same building of a local chemical plant, where they mixed formalin, methanol, and hydrochloric acid in two 3800-liter kettles to produce CMME. During the process fumes were often visible. To check for losses, the

> lids on the kettles were raised several times during each shift. The employees considered it a good day if the entire building had to be evacuated only three or four times per eight-hour shift because of noxious fumes.

The article went on to say that when a retrospective investigation of the fourteen cases was made, by an examination of hospital records and autopsy results and by consultation with family physicians, it was determined that all the men had indeed developed lung cancer; that their age at diagnosis ranged from thirty-three to fifty-five; that the exposure of thirteen of them to CMME ranged from three to fourteen years; and that thirteen of them had died within twenty months of diagnosis. The article concluded that the data "strongly suggest that an industrial hazard is associated with CMME."

At that point, I put down the article, telephoned Samuels in Washington, and asked him if he knew which plant it referred to.

"Sure," he said. "It's the Rohm and Haas factory in Philadelphia."

During the winter of 1973, some significant changes of personnel took place in the Department of Labor, which were said to have been instituted under the direction of Charles W. Colson, special counsel to the President. In January, George Guenther's resignation as Assistant Secretary of Labor and director of the Occupational Safety and Health Administration was accepted by the White House—presumably because his performance had received some highly publicized negative reactions from organized labor. Guenther was replaced by John H. Stender, a former vice-president of the International Brotherhood of Boilermakers, Ironshipbuilders, Blacksmiths, Forgers, and Helpers, who was also a former Republican state senator in Washington. Then, in February, James D. Hodgson, who had resigned as Secretary of Labor to return to the Lockheed Aircraft Corporation as senior vice-president for corporate relations, was replaced by Peter J. Brennan, the president of the

New York City Building and Construction Trades Council, who had organized the counterdemonstration of hardhat construction workers that disrupted the student peace rally at City Hall in New York on May 8, 1970. Early in March, I learned from Samuels that President Nixon's Office of Management and Budget had plans for a reorganization of federal agencies which included a scheme for dismantling large parts of the Department of Health, Education, and Welfare and the Department of Labor, and merging them into a Department of Economic Affairs. "There is considerable speculation that this shuffle may be attempted without the consent of Congress," Samuels said grimly. "If it takes place, it will effectively do away with the separate roles that were envisioned for the Occupational Safety and Health Administration and NIOSH by the Congress when it wrote and passed the Occupational Safety and Health Act. It will also help to sweep under the rug the whole occupational-health scandal we've been trying to expose."

Since then, it seems, whatever plans the Office of Management and Budget had in mind for HEW and the Department of Labor have been held in abeyance pending a resolution of the Watergate crisis. In May, however, NIOSH was transferred from HEW's Health Services and Mental Health Administration (which was dissolved) to its Center for Disease Control, as part of a decision by HEW to reduce manpower and reorganize health programs. Some people saw this as a further downgrading of NIOSH, others as an attempt by HEW to preserve some semblance of a preventive-medicine program. In any case, Civil Service regulations regarding seniority deprived NIOSH of more than fifty of its six hundred and fifty employees. Since it already lacked sufficient funds and manpower to keep up with increased demands for research training, industry-wide safety-and-health studies, documents on criteria, and health-hazard evaluations, there were predictions of serious work backlogs and an attendant lowering of morale. In June, the confusion surrounding the future of NIOSH and its operations was compounded by Secretary Brennan, who requested support from the Office of

Management and Budget for merging NIOSH with the Department of Labor. Commenting on the general situation in the middle of the month, Dr. Marcus Key, the NIOSH director, noted that all its activities would have to be curtailed, and that "the health and safety of the American worker is not going to be protected as much"—a rather chilling prediction in view of the fact that only the previous year Secretary Richardson had estimated that occupational disease killed a hundred thousand American workers each year.

Meanwhile, on April 4th, a hearing on the petition filed by the AFL-CIO's Industrial Union Department and five trade unions for review of the Secretary of Labor's five-fiber standard for exposure to asbestos dust had been held in the United States Court of Appeals for the District of Columbia Circuit. In a brief for the petitioners, the Industrial Union Department argued that the Occupational Safety and Health Act had clearly required the Secretary to set a standard that would adequately protect the health of workers. The Industrial Union Department's case that the Secretary had not done so was based largely upon the integrity of the NIOSH document on criteria for asbestos, which had recommended that a two-fiber standard go into effect within two years. However, in the brief for the respondent—former Secretary Hodgson—there were two documents of startling origin which defended the Secretary's action in declaring a five-fiber standard for four more years. The first of these was a nineteen-page single-spaced critique of the conclusions and recommendations of the NIOSH criteria document, which had been submitted to Dr. Charles Powell, an assistant director of NIOSH, on January 11, 1972, by Dr. George Wright, a longtime paid consultant of Johns-Manville, who was then head of medical research at St. Luke's Hospital in Cleveland, and who testified later that year in behalf of Johns-Manville (and against the proposed two-fiber standard) at the Department of Labor's public hearings. Curiously, Dr. Wright's letter to Dr. Powell was never submitted as part of the public record that the Department of Labor was required to compile in order to establish a permanent standard for asbestos but was only later submitted by Dr. Powell as an extra-record statement solicited by Assistant Secretary of Labor Guenther in order to justify the Department's controversial decision to delay the imposition of a two-fiber standard for four years.

The second letter in support of the government brief was an extra-record memorandum sent to Guenther by Dr. Key, on May 30, 1972—three days before Guenther signed the contested standard for asbestos—which, in effect, disavowed critical portions of the criteria document that had been prepared by Dr. Key's own staff, and under his direction. Dr. Key's memorandum to Guenther ended, "In summary, if your hearings and feasibility study indicate that a two-fiber-per-cubic-centimeter level is not achievable until four years hence, I would accept this as a reasonable health standard." Dr. Key added three safeguards to his acceptance of the higher standard, including provisions that new plant construction be designed to meet a two-fiber level; that plants and operations that had already achieved a two-fiber level be required to maintain it; and that an anti-smoking campaign be required for workers exposed to asbestos. As it turned out, none of these safeguards were incorporated in the standard promulgated by the Department of Labor. However, the government did make liberal use of Dr. Key's memo ten months later, in its brief defending the Secretary of Labor from the petition brought against him by the unions.

Thus did the director and an assistant director of NIOSH take the astonishing step of compromising *in camera* their own asbestos-criteria document and its recommended two-fiber standard, which Secretary Richardson had assured President Nixon, less than a month before Dr. Key wrote his memo, was designed "to protect against asbestosis and asbestos-induced cancer . . . and to be attainable with existing technology." Moreover, as justification for this extraordinary administrative act, Dr. Powell chose to rely on an opinion provided by Dr. Wright, the Johns-Manville consultant (the company was later an *amicus curiae* on the government side), and Dr. Key professed no qualms about feasibility data gathered by Arthur D. Little, Inc., a firm that had relied mainly upon information solicited from the asbestos industry, and from a group of medical doctors the vast majority of whom had conducted research on asbestos disease that was financially supported by the industry. No one knows for sure what could have motivated Dr. Key to sign this extraordinary memorandum, which turned out to have been written by Dr. Powell, but many people assume that he was reacting to pressure from his confreres in the Department of Labor, with which NIOSH might be merged. As for the Industrial Union Department's petition for review of the five-fiber standard, the United States Court of Appeals refused to make a clear-cut decision, but returned it to the Secretary of Labor for his review on April 15, 1974. Many observers feel that the fact that the integrity of the criteria document on asbestos was questioned by its primary author had dealt a severe blow to the Industrial Union Department's case.

By the late spring of 1973, I had long since come to the conclusion that there would be no quick end to the resistance of the medical-industrial complex to action that would ameliorate the plight of workers exposed to toxic substances, or, for that matter, to the capacity of many key government officials to react with timidity and deceit whenever they were required to make decisions regarding occupational-health problems which might run counter to the interests of the corporate giants that had been supplying money and manpower to political administrations for decades. One of the chief difficulties in overcoming the traditional business-as-usual approach to industrial disease was simply that public opinion could not easily be aroused against the delayed carnage occurring in the workplace. In short, unlike the casualty figures of the Vietnam war, which for years had been reported weekly from a specific geographical area, specific or dramatic reporting of casualties from industrial disease could never be provided, for they were occurring years after the onset of exposure to toxic substances and among men and women who had been working in literally tens of thousands of shops and factories in every state of the union. For example, who would be likely to remember forty years from now that several hundred men in the hill country of East Texas who had died of asbestosis, lung cancer, gastrointestinal cancer, and mesothelioma had once been employed at a small insulation plant in Tyler owned by Pittsburgh Corning?

It therefore appeared that only an unusual disaster—a drama of vast magnitude—was likely to evoke the kind of public outrage that, as in the case of the war, would demand an end to the unnecessary slaughter. As things turned out, newspapers and television stations around the country were carrying stories about the possibility of just such a disaster—one that could affect the hundred thousand citizens of Duluth, Minnesota. Whether or not the Duluth situation proves to be truly catastrophic, it certainly serves to illustrate how a potential disaster can evolve without warning, and to suggest how other such disasters are bound to occur in the future if steps are not taken to curb industry's indiscriminate use of the environment as a private sewer. The Duluth story broke in the middle of June, when the Environmental Protection Agency announced that the public water system of the city, which derives its supply from Lake Superior, contained grossly excessive amounts of asbestos-like material. Upon further analysis, this turned out to be similar to amosite asbestos—the type that was handled in the Paterson and Tyler factories. The contamination of Lake Superior with asbestos had evidently begun in the middle fifties, when a new process was developed for extracting iron from the taconite-ore deposits in the nearby Mesabi Range. This process consisted in crushing the ore, grinding it in water to a fine muddy sand, and magnetically separating the iron from the wet slurry. In 1956, the Reserve Mining Company—a three-hundred-and-forty-million-dollar subsidiary of ARMCO and Republic Steel—began using the new process at its plant in Silver Bay, a town on the lake about fifty miles northeast of Duluth, and also began dumping, through pipes and chutes, thousands of tons of pulverized waste tailings into the lake each day. (At the same time, large quantities of mineral dust began to be released into the ambient air through the smokestacks of the plant.) By the middle sixties, when the company was dumping sixty-seven thousand tons of waste tailings into the lake each day, a large green stain almost twenty square miles in area was spreading over the lake in the vicinity of Silver Bay—a stain that was thought for a long time to be composed of algae whose growth was stimulated by the dumping of the tailings. In 1969, concerned about the condition of the water in the lake, the State of Minnesota went to court seeking to enjoin Reserve from the practice of dumping. In the course of the proceedings, it was established that the green stain resulted not from algae but from a light-scattering effect that was caused by sunlight shining on suspended particles. Then, in December of 1972, suspecting that asbestos might be present in the taconite ore mined for production, officials of the Minnesota Pollution Control Agency commissioned two geologists—Stephen Burrell, of the University of Wisconsin, and James Stout, of the University of Minnesota—to undertake a study of the situation. Over the next seven months, Burrell took samples of ore at Reserve's Peter Mitchell Mine, and when Stout analyzed them, asbestos-type material turned out to constitute about twenty per cent of the total. After the iron was separated from the ore, the waste tailings that were left contained an even greater percentage of this material. Thus, since the dumping had begun, it could be estimated that the waters of the lake had been polluted not only with about two hundred million tons of ore refuse but with tens of millions of tons of asbestos-type minerals. Small wonder that when the Environmental Protection Agency—which had instituted its own suit against Reserve in February of 1972—collected samples of the public water supply of Duluth and sent them to the Environmental Sciences Laboratory of the Mount Sinai School of Medicine, in June, to be analyzed they were found to contain approximately a hundred times as much asbestos-type fiber by weight per liter as any other water samples that had ever been analyzed there.

During May, preliminary results of the Burrell-Stout study were reported to officials of the EPA, in Washington, who, after mulling over their implications and the possible effects of their public disclosure (including the possibility of panic among the citizens of Duluth), furnished the information to the court, thus adding a new dimension to the original suit. At the same time, the EPA asked Dr. Irving Selikoff and Dr. E. Cuyler Hammond to evaluate within sixty days the possibility of adverse effects upon the health of the citizens of Duluth. Shortly thereafter, United States District Court Judge Miles W. Lord, who was scheduled to hear the case against Reserve, requested Dr. Selikoff and Dr. Hammond to conduct a preliminary study and report their findings to him within two weeks. Since no studies had ever been performed on people whose exposure to asbestos was purely by ingestion, Dr. Selikoff and Dr. Hammond, and Dr. William J. Nicholson, associate professor of community medicine at the Mount Sinai School of Medicine, and also a member of the staff of its Environmental Sciences Laboratory, undertook to determine the presence of asbestos fibers in the tissues of such people by comparing autopsy material from people who had lived all their lives in Duluth with autopsy material from asbestos workers who had been employed at the old Union Asbestos & Rubber Company's factory in Paterson. Because workers at that plant—and other asbestos workers as well—were known to have incurred three times as much gastrointestinal cancer as people in the general population, the project appeared to be an urgent one.

Toward the end of June, Dr. Selikoff, Dr. Hammond, and Dr. Nicholson informed Judge Lord that because their initial studies showed that the autopsy material from Duluth had been prepared with formalin which had been diluted with water from the Duluth water supply, the material was contaminated with asbestos to start with, and thus made its analysis very much more complicated. For this reason, a proper evaluation of the potential health hazard to the residents of Duluth from drinking water contaminated with asbestos could not be made on short notice but would take many months of painstaking study and evaluation. As a result, the study continued through the summer and fall.

Meanwhile, the suits that had been brought against Reserve by Minnesota and the EPA had been merged, and went to trial on August 1st in United States District Court, with Judge Lord presiding, and with the Justice Department (acting in behalf of the EPA, and joined in its action by Minnesota, Wisconsin, and Michigan, as well as by five environmental groups) contending that Reserve's discharge of taconite waste into Lake Superior posed a threat to public health and should be halted. Because of the complexity of the issues involved, most of the government's key witnesses were not called to the stand until the early part of September. One of the first of these was Dr. Nicholson, who testified that his analysis of water samples from Duluth showed that they contained very large amounts of amphibole minerals. (Amphiboles are a complex group of silicate minerals that chiefly contain magnesium, silicon, and iron. Included in the group are five different varieties of asbestos, including crocidolite and amosite, both of which are known to have caused asbestosis, lung cancer, mesothelioma, and cancer of the gastrointestinal tract in workers who inhaled their fibers.) When asked how many amphibole fibers someone might ingest from drinking a quart of water from the public water supply of Duluth, Dr. Nicholson testified that his analysis showed that a quart of Duluth water could contain anywhere from twenty million to a hundred million such fibers. Later in his testimony, Dr. Nicholson indicated that Reserve's operations might also constitute a serious air-pollution hazard, since his analysis of air samples taken in the vicinity of the company's plant in Silver Bay showed that they contained concentrations as high as eleven million amphibole fibers per cubic meter of air. (Anyone breathing such air could inhale more than one hundred million fibers in twenty-four hours.) Basing his conclusions on the air and water samples he had analyzed, on his research on people who had been occupationally exposed to asbestos, and on his knowledge of other research into the biological effects of asbestos which showed that asbestos-induced cancers usually took twenty years or more to develop, Dr. Nicholson stated that the situation resulting from Reserve's discharge of taconite waste constituted a serious public-health hazard. "If we waited until we saw the bodies in the street, we would then be certain that there would be another thirty or forty years of mortality experience before us," Dr. Nicholson said. "We would have built up a backlog of disease over which we would have little control."

Dr. Nicholson was followed on the witness stand by Dr. Harold L. Stewart, who had been engaged in cancer research for the United States Public Health Service's National Cancer Institute since 1939, and who had been chief of pathology at the Institute from 1954 until 1969, when he retired. Dr. Stewart testified that in his opinion the amphibole fibers in the Duluth water supply constituted a carcinogen. "You give it to the infants," he said. "You give it to young children. This is a captive population. They not only ingest the water, it's virtually a food additive. Everything that's cooked is cooked in [asbestiform minerals]. All the sheets and the pillowcases and the clothes are laundered in the asbestos water. It must be in the atmosphere. It must be floating around. The dryer that dries the clothes in the cellar must blow this out somewhere, I would assume. It's a carcinogen introduced through the domestic water supply into the homes of people." Toward the end of his direct testimony, Dr. Stewart said that anybody who permits such a situation "must realize that he's condemning people to exposure to a carcinogen that may take their lives and probably will."

After Dr. Stewart's appearance, the government called Dr. Arthur M. Langer, the chief mineralogist and head of the physical sciences section at the Mount Sinai Environmental Sciences Laboratory, who testified that his analysis of fibers in samples of the Duluth water supply showed that half of them were of the amphibole variety, that approximately two per cent were identical with amosite asbestos, and that between four and five per cent were consistent with amosite. Dr. Langer also stated that almost half of the fibers found in the air samples taken in Silver Bay were either identical to or consistent with amosite asbestos in their chemical composition. Dr. Langer was followed to the stand by Dr. Joseph K. Wagoner, director of NIOSH's Division of Field Studies and Clinical Investigations. A considerable portion of Dr. Wagoner's testimony was devoted to the excessive amosite-asbestos dust concentrations that Dr. Johnson and others from the Division of Field Studies had found during their survey of Pittsburgh Corning's Tyler plant, in October of 1971, and to the critical occupational-health situation that had existed there. During this testimony, it was brought out that previous surveys of the Tyler plant, which showed gross abuses of good industrial-hygiene practices at the factory as far back as 1967, had been conducted under the direction of Dr. Wagoner's predecessor at the division, Dr. Lewis J. Cralley; that Dr. Cralley had reported the results of his findings to Pittsburgh Corning but not, insofar as Wagoner knew, to any federal regulatory agency; and that Dr. Cralley was at present a consultant for Reserve.

Dr. Wagoner was followed on the witness stand by Dr. Selikoff, who, during four days of testimony, proceeded to review the history of asbestos disease, the results of the major studies of the biological effects of asbestos which had been carried out during the past forty years, and the results of the studies he and Dr. Hammond had conducted showing the disastrous mortality experience of asbestos workers at the Tyler plant and at its predecessor, the Union Asbestos & Rubber Company's factory in Paterson. Dr. Selikoff went on to describe some preliminary findings of a study he is conducting of the relatives of men who had worked at the Paterson plant. He told the court that upon examining a hundred and fifteen people who had lived in the same house with workers at the plant (and who could, therefore, have been exposed to asbestos dust brought home on the workers' clothes), he found that thirty-nine per cent showed X-ray abnormalities, most of them in the form of lung scarring typically found among people who are occupationally exposed to asbestos. Dr. Selikoff also testified that he felt it was highly probable that ingestion of asbestos fibers was responsible for the fact that he had found a threefold increase in cancer of the gastrointestinal tract among the asbestos workers he had studied.

In attempting to deny the government's charges, Reserve contended that the fibers found in the Duluth water supply were not caused by its disposal of taconite waste into Lake Superior, and that, in any case, the fibers were not of the same chemical type that had been found to cause cancer of the lungs and gastrointestinal tract among asbestos workers. However, Dr. Selikoff testified that he strongly believed, on the basis of his experience and upon evaluation of all the data he had studied, that cancer was induced by the "size and shape of the particles rather than their exact chemical composition." Later, when asked if he had an opinion as to whether or not the presence of amphibole fibers in the Duluth water system constituted a health hazard, Dr. Selikoff replied that he thought it posed a distinct health hazard to the population of Duluth and to other populations drinking or using such water. "We will not know whether or not these particular circumstances will cause cancer until another twenty-five to thirty-five years have passed," Dr. Selikoff stated. "This is in my opinion a form of Russian roulette, and I don't know where the bullet is located." Upon further questioning, Dr. Selikoff testified that asbestos levels measured in the air at Silver Bay, less than half a mile from two schools, were about ten times as great as asbestos levels measured near sites where asbestos insulation had been used in building construction in New York City (a practice banned by the city as a health hazard in February of 1972); that the asbestos levels in Silver Bay were about the highest environmental levels he had ever seen; and that people who might be breathing air containing such concentrations were, in his opinion, risking the development of mesothelioma.

At Judge Lord's request, a tentative list of witnesses whom the lawyers for Reserve intended to call to the stand was furnished on September 28th. It contained the names and affiliations of sixty-four men, who were listed according to ten different categories of information about which they could be expected to testify. Some of the names were familiar. They included Dr. Cralley and Robert G. Keenan, of Clayton & Associates; Dr. Paul Gross; Dr. Stephen Holmes, now chief health physicist for the Turner Brothers Asbestos Company; and Dr. John Corbett McDonald, of McGill University—all of whom had figured in the riddle about the medical-industrial complex. Also listed were Dr. Donald W. Meals and three other members of Arthur D. Little, including Dr. Charles J. Kensler, the firm's senior vice-president in charge of operations; Dr. Leonard G. Bristol, director of Immunobiological Research Laboratories of the Trudeau Institute, Inc., at Saranac Lake, which Johns-Manville has helped support for many years; and Dr. Hans Weill, professor of medicine in the pulmonary-diseases section of the Tulane University School of Medicine, whose study of asbestosis in workers at a Johns-Manville cement-products plant at Marrero, Louisiana, had been financially supported by the Quebec Asbestos Mining Association, of which Johns-Manville is a leading member.

Whatever the outcome of the trial, all those involved in the Duluth affair, and especially the city's hundred thousand residents, are keeping their fingers crossed. On the face of it, the predicament would seem to be an incredibly absurd and hapless one for the people of any city in this country—the most highly developed and technologically expert nation of the world—to be in. The major steel companies are anxious about it, because many of them have mines and mills in the Mesabi Range, and the health of thousands of their workers may have been seriously jeopardized through exposure to asbestos dust. Most anxious of all, however, are ARMCO and Republic Steel, for, as co-owners of Reserve, they might, if worst comes to worst for the people of Duluth, be sued under Minnesota law. As it happens, one of the chief medical consultants for Republic Steel these days is none other than Dr. George Wright. Dr. Wright's value to Republic Steel in the Duluth affair may prove to be limited by the fact that while in the service of Johns-Manville—the world's largest producer of chrysotile asbestos—he testified at the Department of Labor's public hearings that amosite, not chrysotile, was, in his opinion, responsible for the grossly excessive incidence of mesothelioma in the study of insulation workers conducted by Dr. Selikoff and Dr. Hammond.

During the summer of 1973, while the situation in Duluth was making headlines across the country, the Department of Labor addressed itself to the business of setting permanent standards for exposure to the fourteen chemical carcinogens for which it had promulgated six-month temporary emergency standards on May 3rd. As in the case of the asbestos standard a year and a half before, the Secretary of Labor convened an advisory committee—the Standards Advisory Committee on Carcinogens—in order to avail himself of expert advice on the problem. The committee was made up of fifteen members drawn from government, labor, the independent medical and scientific community, and industry, and in August, it recommended to Secretary Brennan that the permanent standards include a strict permit system, so that no company could manufacture or use any of the fourteen carcinogens without first demonstrating its ability to handle the chemical in a manner that would avoid detectable exposure to any worker. This recommendation came about largely because of evidence presented to the advisory committee by Dr. Wagoner's Division of Field Studies and Clinical Investigations, which demonstrated that literally hundreds of factories across the land were manufacturing and using the chemicals with either grossly inadequate or, in some cases, no controls for the protection of workers.

In September, when the Department of Labor's Occupational Safety and Health Administration held public hearings on the proposed standards, in Washington, D.C., the recommendation of the advisory committee was discounted by Leo Teplow, a special consultant for Organization Resources Counselors, Inc. —a concern representing forty major corporations, many of which either manufacture or use one or more of the carcinogens involved. Testifying at the hearings on September 11th, Teplow declared that the recommendation merited very little attention, and added, "This is a subject for consideration by people of far more generalized experience than those who were primarily toxicologists and hygienists that composed this committee."

On September 14th, Teplow's statements were contradicted by Dr. Key, of NIOSH, who on this occasion took a strong and forthright stand on the matter of occupational exposure to known carcinogens. "It is the firm conviction of NIOSH that no evidence exists to scientifically conclude that any one of the fourteen agents under consideration is not carcinogenic in man," Dr. Key testified. "I make this presentation as a concerned public-health official in the spirit of fostering a better understanding of occupational carcinogenesis . . . and also in the spirit of encouraging a prudent approach to standards setting, so that we can profit from those lessons learned from experience with beta-naphthylamine and bis-chloromethyl ether. Such lessons, although advancing the state of knowledge as far as experimental versus human carcinogenicity is concerned, were unfortunately done at the expense of the health of the American worker."

On the same day, Samuels, of the Industrial Union Department, attacked the timing of the public hearings. "This hearing could have been held more than two years ago, when our department and affiliates first questioned the exclusion of known carcinogens from the interim standards promulgated under the Occupational Safety and Health Act," Samuels testified. "The consequence is that, as a result of two years of unjustifiable exposure to carcinogenic agents, regardless of anything done now, hundreds and perhaps thousands of men and women can be expected to experience agonizing death from cancer in the next two decades. Sir, in a truly civilized society we would hold personally responsible those who participated in this crime, both the callous political creatures and the cancer peddlers who bartered moral and statutory obligations. In a just society they would now be undergoing rehabilitation in a penal institution. Instead, they walk freely—some of them are or have been in this room—as if evil is its own reward."

Although Secretary Brennan and Assistant Secretary Stender were required by law to announce the promulgation of permanent standards for the fourteen carcinogens by November 3rd, that deadline passed with no action taken. However, on November 8th, Gerard F. Scannell, the director of the Occupational Safety and Health Administration's Office of Standards, informed Samuels that the new standards would not include a permit system, which had been recommended by Secretary Brennan's advisory committee, by NIOSH, and by the unions, so that the standards could be effectively enforced under the 1970 Act. This proved to be the case, for on January 29, 1974, the Secretary promulgated permanent standards for the fourteen carcinogens which did not include a permanent system, or provide for monitoring the air of the workplace, or for medical surveillance of the workers. Thus, once again, as in the case of the asbestos standard, a Secretary of Labor disowned the recommendations of the members of his own advisory committee and of occupational-health experts from NIOSH. In this instance, he even disowned the advice of the staff of the President's Council on Environmental Quality, who also strongly urged the necessity of including a permit system in the new standards. Indeed, continuing the long business-as-usual approach of the federal government toward the crucial problem of industrial disease, he set standards for known cancer-producing substances which failed to adequately protect the health and lives of workers.

Meanwhile, in the autumn of 1973, there were some new developments concerning Pittsburgh Corning's Tyler plant, which had provided the starting point for my twenty-one-month investigation of the medical-industrial complex. In October, I received a copy of a report entitled "Tyler Asbestos Workers Study"—a project designed to conduct a medical followup of the Tyler workers in the coming years—which had been drawn up by a newly formed, nonprofit organization called the Texas Chest Foundation, with headquarters at the Texas State Department of Health's East Texas Chest Hospital, in Tyler. The idea for the project had been initiated by Dr. Johnson late in the summer of 1972, and it had been developed since then by him and by Dr. Wagoner; by Richard A. Lemen, an epidemiologist in the Division of Field Studies and Clinical Investigations; by Dr. George A. Hurst, the superintendent of the East Texas Chest Hospital; and by Dr. Selikoff. Upon making inquiries, I learned that these men had met in May and June of 1973 with Dr. Michael B. Sporn, the chief of the lung-cancer branch of the Department of Health, Education, and Welfare's National Cancer Institute, for the purpose of engendering interest in and funding for the project. I also learned that during the late spring and summer Lemen had managed to trace the whereabouts of six hundred and ninety-two of the eight hundred and ninety-five men who had been employed at the Tyler plant from November of 1954, when it opened, until February of 1972, when Pittsburgh Corning shut it down. In addition, I found out that on June 18th, Dr. Johnson had written Dr. Grant, medical consultant to Pittsburgh Corning, informing him of the proposed study and of the National Cancer Institute's interest in it.

When I talked with Lemen, he told me that in August of 1973 he decided to revisit the site of the old Tyler plant, where the Imperial American Company now manufactures lawn furniture. "I had not been back to Tyler since January of 1972, when Steven Wodka and I visited the dumps where Pittsburgh Corning was disposing of asbestos waste," he told me. "Imagine my surprise when I discovered that the soil with which the company had later covered them over had eroded, and that loose wads of asbestos fiber were lying everywhere on the ground." Lemen went on to tell me that he had brought the matter to the attention of officials of the Texas State Department of Health and of the EPA. I later learned that Pittsburgh Corning had been directed by the EPA this past September to remedy the situation. The company then proceeded to re-cover the dumps with two feet of soil and with four inches of topsoil, and to seed them with grass, at a total cost of some sixty thousand dollars. The prudence of the EPA's action was reinforced by an article that appeared in August of 1973 in the *Annals of Occupational Hygiene.* The article was entitled "Asbestos in the Work Place and the Community"; it was written by Dr. Muriel L. Newhouse, of the London School of Hygiene and Tropical Medicine; and it included the description of "a patient suffering from a peritoneal mesothelioma [who] recalled playing with handfuls of asbestos on waste ground near a factory as a small boy, some forty years before he developed his tumour."

As for the "Tyler Asbestos Workers Study," which was approved for funding by the National Cancer Institute in the spring of 1974 (and to which, as yet, Pittsburgh Corning has not offered to make any contribution), it says, under the heading "Definition of the Problem":

> In 1972, approximately eight hundred and seventy-five employees of a Tyler asbestos plant completed massive exposure to inhaled amosite asbestos fibers. Approximately two hundred and sixty of these employees will develop cancer, mainly bronchogenic carcinoma, and die from this disease unless some type of successful intervention is undertaken.

At present, of course, no truly successful intervention to combat the development of lung cancer is known to the medical and scientific community, but it is at least a hopeful sign that because the National Cancer Institute has funded the project the Tyler workers will not be wholly abandoned to the awful peril they have been exposed to (as have so many other asbestos workers in the United States), and that if in the next few years some kind of successful intervention is discovered some of these men who might otherwise die unnecessarily early deaths may be saved.

Shortly before the Duluth story broke into the headlines, I received an announcement of the National Symposium on Occupational Safety and Health, which was to be held at the Carnegie Institution, in Washington, D.C., from June 4th to June 6th. Among the listed sponsors of the symposium were the Division of Industrial and Engineering Chemistry, the American Chemical Society, the Manufacturing Chemists Association, the National Safety Council's Chemical Section, NIOSH, and the Occupational Safety and Health Administration. According to the announcement, Session Six of the symposium, entitled "The Human Factor," would be co-chaired by Dr. Grant, who was scheduled to deliver some introductory remarks.

Several months later, I received a copy of the October 3, 1973 issue of *Chemical Week*, a publication put out by McGraw-Hill. It contained an article entitled "Health Programs Need First Aid," which described a meeting of the Manufacturing Chemists Association, held in Atlanta during September. The article said that "O.S.H.A., now in its third year, had issued its first permanent health standards—for asbestos," and that "ten or more substances will have permanent standards by the end of the year, according to Lee Grant, medical director of PPG Industries, who moderated the discussion." The article went on to state that "Grant says O.S.H.A. will define a 'healthful environment' and take other steps concerning methods and sampling procedures in the work environment, medical examinations for employees, and prescribed work practices."

Although I was unable to attend either the National Symposium on Occupational Safety and Health or the Manufacturing Chemists Association meeting, I had had a chance to see Dr. Grant preside over a similar gathering almost a year before. The occasion was the One Hundredth Annual Meeting of the American Public Health Association, which was held in Atlantic City from November 12th through November 16th of 1972. At that time, nearly eight months had passed since James M. Bierer, the president of Pittsburgh Corning, telephoned me to say that he could not give me permission to talk with Dr. Grant, or with any other Pittsburgh Corning employee, about the Tyler plant. During those eight months, I had gathered a good deal of information about Dr. Grant, and it had made me all the more curious to meet him. Therefore, when I received a copy of the official program of the American Public Health Association's meeting and saw that he was scheduled to preside at two of its sessions, I decided to attend, in the hope of talking with him. I arrived in Atlantic City on Sunday, November 12th, and went directly to a meeting of the American College of Preventive Medicine, which was being held in Ballrooms A and B of the Holiday Inn and was being presided over by Dr. Grant in his capacity as the college's outgoing president.

Upon reaching the doorway of the ballroom, I could see half a dozen people sitting on a platform at the front of the room, about twenty yards away. I asked a young man standing in the doorway if he knew which of them was Dr. Grant, and he pointed out the man sitting at far stage right—a lean blond man, with deep-set eyes and a prominent nose, who looked to be in his middle forties. However, when the session ended, a few minutes later, and I advanced across the ballroom floor toward the platform, I realized that Dr. Grant was at least ten years older than I had thought, and that it was his blond hair—parted and combed in the flat collegiate style of twenty years ago—that made him appear younger at a distance. When I reached the platform, I introduced myself to Dr. Grant, who was standing above me, at approximately knee-to-eyeball level, and reminded him that I had telephoned him in March in the hope of being able to talk with him about the Tyler plant. I also reminded him that at the time he had referred me to Bierer, who had subsequently refused to give me permission to talk with any employee of the company.

"I wonder if there might be some time in the next day or two that I could talk with you," I said.

Dr. Grant appeared to hesitate. Then, glancing quickly over the ballroom, which was emptying, he shook his head. "I'm afraid I can't," he replied. "In this instance, it's a question of the patient's rights."

For a moment, I thought I had not heard him correctly. Then it dawned on me that he was talking about the company. "Do you mean Pittsburgh Corning?" I said.

"Why, yes," Dr. Grant replied. "If they don't want me to talk with you, there's nothing I can do."

"But isn't the patient all those men who worked in the Tyler plant?" I asked.

Dr. Grant straightened up and looked down at me from his full height on the platform. "Well, in the larger sense, of course, that's probably true," he replied. "And now, if you'll excuse me, I have some business to attend to."

I stood at the platform and watched Dr. Grant, who, as he moved away, put a cigar in his mouth, lit it, and exhaled a cloud of smoke into the air. A moment later, I saw him throw an arm in greeting around the shoulders of a colleague. Then I turned away, and found myself looking straight into the face of Anthony Mazzocchi, who, as it turned out, had been invited to speak at one of the convention's sessions on occupational health.

"Did you hear that?" I asked him. "Did you hear what he said?"

For a long time, Mazzocchi looked at me without a trace of expression on his face. Then, very slowly, he nodded his head up and down. And then, just as slowly, he shook it from side to side.

EPILOGUE

A Pandora's Box

JANUARY 1974

A $100 million class action lawsuit, the largest in recent history, was filed Wednesday in Tyler's U.S. District Court on behalf of former employes of the Pittsburgh Corning Corporation in Tyler.

The suit alleges negligence on the part of Pittsburgh Corning Corporation in exposing employes to "asbestos fibers in extremely dangerous concentrations," causing them to "suffer from various stages of asbestosis and-or lung cancer and-or other pulmonary diseases."

One plaintiff, Jessie Mae Thomas "is the surviving widow of Robert T. Thomas, deceased, a former employe of Pittsburgh Corning who has died as a result of such exposure and subsequent effects therefrom."

Plaintiffs in the suit include Herman Yandle, Arthur Bearden, J. C. Yandle, Harold Spencer, Rufus J. Lee, Hubert T. Thomas and Jessie Mae Thomas, "On Behalf of Themselves and All Other Similarly Situated."

Defendants are PPG Industries, Inc., Dr. Lee Grant, Corning Glass Works, Inc., The Industrial Health Foundation, Inc., The Asbestos Textile Institute, Pittsburgh Corning Corporation and John Doe I-X.

— Carol Paar in the *Tyler Morning Telegraph*, January 3.

AUSTIN (UPI) — A Texas legislator said Wednesday it is much better "to have a little bit of crud in our lungs" than to allow citizens to file lawsuits every time they think someone is violating antipollution laws.

"I don't need some bunch of dogooder nuts telling me what's good to breathe," Rep. Billy Williamson, D-Tyler, said. "And I don't want a bunch of environmentalists and communists telling me what's good for my life and family."

Williamson urged a constitutional convention committee to reject a proposal that would allow citizens to sue state agencies and public officials that do not properly administer state environmental laws.

"I think we are all willing to have a little bit of crud in our lungs and a full stomach rather than a whole lot of clean air and nothing to eat," Williamson said. "I am concerned with my right as a citizen not to be sued every time I turn around by some nut."

— from *The Tyler Courier-Times*, January 24.

AKRON, Ohio — B. F. Goodrich Co., in a move that sent ripples through the plastics and chemicals industries said it is investigating the possibility that the cancer deaths of three workers at a plant in Louisville, Ky., may be job-related.

Goodrich said that the workers affected were in its polyvinyl chloride operations, and that they died of angiosarcoma of the liver, a rare form of liver cancer. . . .

Word that the deaths might be job-releated spread quickly throughout several industries. Said Louis Beliczky, an industrial hygienist on the staff of the United Rubber Workers union, which has many members involved in vinyl operations at various companies: "I look at this as the dropping of a bomb. It opens up the possibility of a whole new occupational disease."

— from *The Wall Street Journal*, January 24.

FEBRUARY

LOUISVILLE (UPI) — An official of the Kentucky Occupational Safety and Health Administration (KOSHA) says the federal government has declared a rare liver cancer, which has taken the lives of four Louisville B. F. Goodrich Co. employees, as a new occupational disease.

"Right now, the only four known cases of this disease are in Louisville, but this thing has tremendous implications nationally and internationally," said Dr. J. Bradford Block, medical consultant for the KOSHA.

— from the *Cincinnati Enquirer*, February 3.

"What I am really saying to you, Mr. Fride, is that unless some dramatic testimony comes into the record in this case you can anticipate prima facie, subject to change, that there will be required an on-land disposal system.

"The only reason that I am saying that to you now is that this case has still three months to go. If at the end of this case I am still as concerned about the public health as I am now, I will consider closing the plant immediately; and what I have been trying to give you for the last three to five months is a three to five-month head start.

"Now, I haven't made up until I heard the evidence—the more evidence I hear the closer I come to a conclusion. Reserve has offered nothing that has weakened or rebutted the testimony of the Government's experts on the health issue insofar as I, as a finder of fact, am concerned.

"As a consequence, I think the options are entirely up to Reserve, whether you are going to start working on something or not. But as I look at what I see as a timetable, I will have to subtract the time that has been spent while you tread water in effect and do nothing, faced with the realities of the public health issue as it now is.

"So that it is up to you now to determine whether you want to put in an alternative plan or if you would just rather have the Court's order. The time has started to run."

— United States District Court Judge Miles W. Lord addressing Edward T. Fride, attorney for the Reserve Mining Company, from the *Court Transcript*, February 5.

"Currently, it is estimated that ten companies with fourteen plants and more than 1,500 production workers are involved in producing vinyl chloride monomer. It is further estimated that some 23 companies and 37 plants employing more than 5,000 production workers are involved in the polymerization of vinyl chloride to polyvinyl chloride. It is difficult to estimate the number of workers employed in converting, molding, and fabricating the polymers into finished products, but it is in the order of tens of thousands working in thousands of plants throughout the country."

—Vernon E. Rose, Acting Institute Director for Research and Standards Development, NIOSH, testifying before an informal fact-finding hearing on possible hazards of vinyl chloride manufacture and use, which was held by the Department of Labor in Washington, D.C., February 15.

". . . there has been evidence of potentially serious disease among workers engaged in vinyl chloride–polyvinyl chloride manufacture for 25 years which has, for reasons not yet well understood, been incompletely appreciated and inadequately approached by medical scientists and by regulatory authorities. In other words, this is not a new problem. What is new is the vigor and extent of our attention to it.

"Thus, for example, acute animal toxicity to vinyl chloride was reported as long ago as 1938 and important chronic toxicity in 1961. Even more, we didn't need the animal studies to warn us that workers could be harmed in VC-PVC manufacture. As far back as 1949, 25 years ago, liver damage was found in 15 of 48 workers in Russia and disease of the liver, skin, and other organs was reported in the next 17 years, not only from Russia but from France and Romania as well. . . .

"Hemangiosarcoma of the liver has in the past been extraordinarily rare. At the Los Angeles County Hospital only one instance was found in 52,000 consecutive autopsies. Our own experience is similar. At one of the institutions of the Mount Sinai Medical Center, Elmhurst General Hospital, no case has been seen despite the very active medical and surgical services of this thousand-bed hospital. In another of our institutions, the Bronx Veterans Administration Hospital, no case has been seen, despite a search of our files, in 30,000 consecutive autopsies and 200,000 surgical specimens since 1941. Yet four cases have been described in one vinyl chloride–polyvinyl chloride plant in the United States in which some 250 workers are employed in the VC-PVC operation. This is an extraordinary increase in incidence."

— Dr. Irving J. Selikoff testifying at the same hearing.

"On 67 animals, [exposed to] 500 ppm [parts per million of vinyl chloride], we got 3 Zymbal glands tumors, 3 nephroblastomas, 7 liver angiosarcomas, 3 tumors of other site and/or type.

"At the exposure of 250 ppm, we got no Zymbal glands tumors, 5 nephroblastomas, 2 liver angiosarcomas, and 4 tumors of other site and/or type. . . .

"And for what concerns the four tumors in the 250 ppm group, we observed 1 Zymbal gland adenoma, 1 intrabdominal angiosarcoma next to spleen, 1 intrathoracic ossifying angiosarcoma, and 1 salivary gland carcinoma."

— Professor Cesare Maltoni, Istituto di Oncologia, Bologna, Italy, testifying about his experiments at the same hearing.

"In summary, the clinical evidence associated with the deaths of four of our employees identifies the need for scientific inquiry beyond the experience of any one company or industry group."

— Anton Vittone, president, B. F. Goodrich Chemical Company, testifying at the hearing.

"Invariably, whenever a new occupational cancer is discovered, it is played down for fear of alarming the workers and the general public. . . .

"The serious national question that is raised and not resolved, because it is either too shocking to contemplate or because the agencies responsible for the protection of the public are reluctant to open this complex Pandora's box, is, in essence, that a national study directed at the industrial environment might uncover a whole series, a succession of occupational cancers, which in turn would implicate in the distribution of these chemicals a larger and larger portion of the population exposed to the risk of cancer. . . ."

— Dr. Thomas F. Mancuso, research professor, University of Pittsburgh Graduate School of Public Health, testifying at the hearing.

"I'd like to insert in the verbal record at this point, to complement the testimony of Dr. Mancuso, the following statement of the Surgeon General dated April 22nd, 1970: 'No level of exposure to a chemical carcinogen should be considered toxicologically insignificant for man. For carcinogenic agents a safe level for man cannot be established by application of our present knowledge. The concept of socially acceptable risk represents a more realistic notion.' "

— Sheldon W. Samuels testifying at the hearing.

"Dow urges OSHA to recognize the voluntary industry action to lower the levels of exposure to vinyl chloride and the lead role of industry in obtaining data on the vinyl chloride problem."

— Dr. V. K. Rowe, research scientist, Dow Chemical USA, testifying at the hearing.

The Health Research Group today petitioned the Food and Drug Administration, the Consumer Product Safety Commission, and the Environmental Protection Agency to ban the use of vinyl chloride in aerosolized cosmetics, including hairsprays, in aerosolized household products, and in pesticide products, since there is substantial evidence that vinyl chloride is a carcinogen. . . .

The Health Research Group also requested that the Food and Drug Administration immediately prohibit the use of polyvinyl chloride (PVC) containers to package any cosmetic product which is capable of leaching detectable amounts of vinyl chloride from its plastic container. In 1973, FDA and industry tests of alcoholic beverages packaged in PVC bottles determined that between 10–20 ppm of vinyl chloride had migrated out of the plastic bottles into the distilled spirits or wine they contained. To the Group's knowledge, no tests have been conducted to determine whether other alcohol-containing or organic solvent-containing liquids (e.g., bath lotion or perfume) can also leach detectable amounts of vinyl chloride from their PVC containers. Moreover, although the FDA proposed banning the use of PVC whiskey bottles in May 1973 because of the potential toxicity of vinyl chloride, the agency has failed to finalize the action.

— from a press release issued by the Public Citizen's Health Research Group, February 21.

The current Federal standard allows workers to breathe as much as 500 parts of vinyl chloride per million parts of air, a level that readily causes liver damage in animals, according to a study published 13 years ago by a Dow Chemical Company scientist.

— Jane E. Brody in *The New York Times,* February 22.

Another case of a rare liver cancer has been found among vinyl chloride workers, this one at a Union Carbide plant in South Charleston, W. Va.

Previously, six cases of the fatal disease, called angiosarcoma of the liver, were found among vinyl chloride workers at the B. F. Goodrich Chemical Company in Louisville, Ky.

— from *The New York Times,* February 22.

Last August, Thomas F. Meade, 22, was stricken with a nerve disease after two years of work at Columbus Coated Fabrics Co., a subsidiary of Borden, Inc., in Columbus, Ohio, that makes vinyl-coated fabrics for luggage and wall coverings. The disease, diagnosed as peripheral neuropathy, left Meade crippled, and today he wears leg braces. He is the most severely affected of some 130 workers at the plant who have suffered symptoms of the disease. The lesser symptoms include weakness and loss of coordination.

When the symptoms became widespread last summer, state health officials tested the nerve responses of the plant's 1,124 employees and told 200 of them to stay at home for two months. Many others stayed home, too, out of fear. By December, most employees had returned, and many who had showed neuropathic symptoms appeared well again.

Meanwhile, Ohio health officials had zeroed in on an ink solvent, methyl butyl ketone, used in applying decorative patterns, as the most likely culprit. While continuing to monitor the plant's air, they banned the use of MBK. The plant's managers have switched to a substitute, improved ventilation, issued respirators to print shop workers, and moved lunch areas away from work areas.

But Corwin Smith, president of Local 487 of the Textile Workers Union of America, which represents the plant's workers, is not satisfied. He says that 26 members are still out with the disease, that eight of them are severely crippled, and that three workers whose condition improved are now sick again. On Feb. 9, the local struck the plant as negotiations collapsed on a new contract; one of the union demands was a "proper work environment." Says Smith: "People here don't know whether they will be crippled in the future or not."

— from *Business Week*, February 23.

Dr. Mitchell Zavon, assistant health commissioner of Cincinnati who announced his March 1 resignation Dec. 18, is going with Ethyl Corp. as medical director. He will have offices at Baton Rouge, La. Dr. Zavon has served with the Cincinnati Health Department for 17 years and for the last three years has also been president of his own occupational consulting firm, Agatha Corp.

— from the *Cincinnati Post*, February 26.

MARCH

During its first eight months, O.S.H.A. inspectors visited 17,743 workplaces. At this rate, it would take them more than 200 years to visit each one once. If the staff were expanded to the original projection of 2,000, it would still take about 50 years to complete the rounds of workplaces.

— Jane E. Brody in *The New York Times*, March 4.

PPG Industries, Inc., and Dr. Lee Grant became the [third and] fourth co-defendants to file answer Monday to a $100 million class action suit filed in January on behalf of former PPG employes in Tyler's U.S. District Court.

Corning Glass Works and Pittsburgh Corning Corporation filed similar answers Monday morning denying allegations charged in the complaint. Corning Glass Works said that "members of the alleged class action have no joint cause of action but at most separate and individual causes of action dependent upon different sets of facts . . . so that plaintiffs do not represent a class with a common cause of action."

Pittsburgh Corning Corporation, in a separate document, filed an answer to the plaintiffs' motion for joinder of parties, also.

The answer filed by PPG stated that "this suit is not properly maintainable as a class action . . . and should be dismissed as such."

PPG said, "the plaintiff's have filed or could file workmen's compensation claims for the injuries and damages made the basis of this lawsuit."

The answer further stated that the plaintiffs are barred from recovery by "assumption of risk, contributory negligence and the Fellow-Servant doctrine," as the answers filed earlier also contend. . . .

The answer filed on behalf of Dr. Lee Grant by his attorney . . . denies allegations set forth in the complaint and uses as defenses the two-year statute of limitations law in Texas and the fact that the plaintiffs were covered by workmen's compensation laws in the state of Texas. . . .

Grant, an employe of PPG Industries, Inc., admits in the answer that he visited the Tyler plant and surveyed the conditions present on one or two occasions, but denies, as charged in the complaint, that he "failed and refused to advise such workers of health hazards and of their own physical problems." Grant requested a trial by jury in his answer, also.

— Carol Paar in the *Tyler Morning Telegraph*, March 5.

WASHINGTON — Over the objections of industry, federal scientists are recommending that worker exposure to vinyl chloride be reduced to amounts too small to be measured in order to avert possible liver cancer.

Federal standards currently limit exposure to vinyl chloride to 500 parts per million in the atmosphere, and industry experts have urged that the limit be set at 50 parts per million.

However, scientists at the National Institute for Occupational Safety and Health are unable "to describe a safe exposure level," Dr. Marcus Key, the institute's director, advised the Labor

Department. Therefore, he added, "We rejected the concept of a threshold limit for vinyl chloride gas in the atmosphere."

— from *The Wall Street Journal*, March 13.

WASHINGTON, March 28 (UPI) — The Environmental Protection Agency announced today that it would ask pesticide makers voluntarily to stop using an aerosol propellant gas, vinyl chloride, that has been linked to a rare form of liver cancer. . . .

The Health Research Group had also asked E.P.A. to announce the brand names of pesticides using vinyl chloride, but the agency indicated it was not yet ready to comply.

— from *The New York Times*, March 29.

APRIL

MINNEAPOLIS, April 2 — The president of the Reserve Mining Company has admitted in Federal District Court here that he is responsible for the company's having withheld vital evidence from the judge and the plaintiff in a complex pollution trial.

The company president, Edward M. Furness, has also accepted responsibility for Reserve's having provided evidence that Federal District Judge Miles Lord said had misled him and greatly prolonged the trial, now in its ninth month. . . .

Judge Lord and the plaintiffs have repeatedly asked Reserve witnesses to disclose whether the company has prepared alternative plans to discharge the waste on land.

Until recently, Reserve steadfastly denied that it had such plans. At the same time, the mining concern had offered in court a plan to modify its method of discharging the tailings into the lake.

However, on March 1 it was revealed in court that Reserve had prepared perhaps four or five plans for disposal of the waste on land. In one instance, the company had prepared such plans as early as 1970. . . .

Mr. Furness testified that he was "familiar" with a plan for disposal on land prepared in 1970 by Reserve engineers. The company president also told Judge Lord that he had given a key Reserve witness authority to testify that such plans did not exist. . . .

The judge then ordered Armco and Republic brought into the case as co-defendants, and he has ordered the two parent corporations to produce all documents they have on the matter of disposal or face heavy fines. The case resumes tomorrow.

— from *The New York Times*, April 3.

WASHINGTON, April 3 (AP) — Clairol, Inc., announced today that it was recalling from the nation's store shelves about 100,000 cans of aerosol hair spray, some containing a chemical recently linked to a rare form of liver cancer.

The cosmetics concern said that the request for a voluntary recall had come from the Food and Drug Administration following reports that at least 10 industrial workers exposed to the chemical, vinyl chloride, had developed angiosarcoma.

The two brands being recalled are Summer Blonde Aerosol Hair Spray and Miss Clairol Aerosol Hair Spray. The concern said that vinyl chloride had been used in the propellant of those two brands for about four years but had been discontinued last summer.

— from *The New York Times*, April 4.

The employees of the B. F. Goodrich Chemical Company who died from angiosarcoma of the liver had an average exposure of approximately 19 years to vinyl chloride, at unknown concentrations, and variable exposures to other volatile chemicals. (TR 93). Some employees of Union Carbide Company and Goodyear Company are also reported in a post-hearing comment from NIOSH dated March 11, 1974, to have had exposure to vinyl chloride and to have died from angiosarcoma of the liver. Finally, autopsies of four deceased employees revealed that liver angiosarcoma tumors were histologically indistinguishable from the angiosarcoma tumors observed in Professor Maltoni's experimental animals. It is concluded therefore, that vinyl chloride is carcinogenic for humans.

We therefore conclude that the present standard for VC should be lowered from a ceiling of 500 ppm to a ceiling of 50 ppm. . . .

— from the Occupational Safety and Health Administration's Emergency Temporary Standard for Exposure to Vinyl Chloride, published in the *Federal Register*, Volume 39, Number 67, April 5.

WASHINGTON, April 4 — The Environmental Protection Agency announced today that it had begun a study to determine what happens to the more than 300 million pounds of vinyl chloride that escapes into the environment each year in the process of manufacturing plastics. . . .

. . . An E.P.A. study of vinyl chloride exposure resulting from the use of an antiseptic room spray containing vinyl chloride indicated that if sprayed for 30 seconds in a typical bathroom, levels of vinyl chloride in the room air would reach 400 parts of the chemical per million parts of air.

This level is more than double that shown to cause cancer in laboratory animals and is far above the emergency standard of 50 parts per million that the Department of Labor has said workers may be exposed to.

— Jane E. Brody in *The New York Times*, April 5.

An industry-sponsored study of vinyl chloride has indicated that when mice are exposed to the amount of the chemical that workers are currently permitted to inhale, the animals develop a rare fatal cancer of the liver that has been found in 12 vinyl chloride workers. . . .

The industry's laboratory finding, presented yesterday to a private briefing for Government and industry officials, raises questions about the safety of the current Federal emergency occupational standard of 50 parts of vinyl chloride per one million parts of air.

— Jane E. Brody in *The New York Times*, April 16.

WASHINGTON, April 17 (AP) — The Food and Drug Administration has announced the recall of 29 more brands of aerosol products containing vinyl chloride. . . .

The products include medicated vaporizers, athlete's foot sprays, hair sprays and wig spray and cleaner.

Earlier, the agency ordered the recall of 22 brands of consumer and professional hair sprays manufactured by Clairol, Inc., and Bonat, Inc., because they also contained vinyl chloride.

— from *The New York Times*, April 18.

WASHINGTON, April 18 (UPI) — The Environmental Protection Agency has named 20 aerosol pesticides containing an ingredient that may cause a rare form of liver cancer. Ten concerns have refused to let the names of other such products be published. . . .

An E.P.A. spokesman said that the 10 concerns refusing to release the brand names had argued that the information was confidential or subject to statutory protection. All will be asked to show cause within 30 days why the names cannot be released, he said.

— from *The New York Times*, April 19.

The Court has been constantly reminded that a curtailment in the discharge may result in a severe economic blow to the people of Silver Bay, Babbit and others who depend on Reserve directly or indirectly for their livelihood. Certainly unemployment in itself can result in an unhealthy situation. At the same time, however, the Court must consider the people downstream from the discharge. Under no circumstances will the Court allow the people of Duluth to be continuously and indefinitely exposed to a known human carcinogen in order that the people in Silver Bay can continue working at their jobs. . . .

Today, April 20, 1974, the chief executive officers of both Armco and Republic have testified that they are unwilling to abate the discharge and bring their operation into compliance with applicable Minnesota regulations in an acceptable manner. . . .

Up until the time of writing this opinion the Court has sought to exhaust every possibility in an effort to find a solution that would alleviate the health threat without a disruption of operation at Silver Bay. Faced with the defendants' intransigence, even in the light of the public health problem, the Court must order an immediate curtailment of the discharge.

THEREFORE, IT IS ORDERED

1) That the discharge from the Reserve Mining Company into Lake Superior be enjoined as of 12:01 A.M., April 21, 1974.

2) That the discharge of amphibole fibers from the Reserve Mining Company into the air be enjoined as of 12:01 A.M., April 21, 1974 until such time as defendants prove to the Court that they are in compliance with all applicable Minnesota Regulations . . .

— from a *Memorandum and Order* issued by United States District Judge Miles W. Lord on April 20.

MINNEAPOLIS, April 20 — . . . The judge called a plan put forth today by C. William Verity, head of Armco and chairman of the Reserve Company board, "absurd" and "preposterous."

In part, the plan called for state and Federal assistance on expenses the company would incur in changing to a land-disposal system, and for permits necessary to the mining operation to be issued for the expected life of the mine.

Judge Lord said that Armco and Republic were two of the nation's wealthiest corporations and that a request for aid was "absurd."

Mr. Verity also asked the court to rule that no health menace existed, and the judge called that "shocking and unbecoming in a court of law." Based on the evidence presented, Judge Lord said, such a request was tantamount to asking a judge "to violate the oath of his office and to disregard the responsibility that he has not only to the people but also to himself."

— William E. Farrell in *The New York Times*, April 21.

SILVER BAY, Minn., April 23 — The 3,100 employees of the Reserve Mining Company returned to their jobs today with a sigh of relief tempered by a continuing uncertainty over the future of the huge ore processing plant and mine that provides them and their families with a living. . . .

. . . late last night, lawyers for the Reserve Company, which is jointly owned by the Armco Steel Corporation and the Republic Steel Corporation, obtained a temporary reprieve from a three-judge panel of the United States Court of Appeals for the Eighth Circuit that was hastily convened in a motel in Springfield, Mo.

The Appeals Court judges allowed the company to resume its operations until May 15, when they will hear Reserve's motion to stay Judge Lord's ruling until an appeal of the entire complex case can be heard.

— William E. Farrell in *The New York Times*, April 24.

MAY

. . . at the Goodyear plant in Niagara Falls, where a team of environmental scientists from Mount Sinai Medical School spent three days doing intensive mass medical examinations of workers, Frank Micale, president of the 300-member Local 8-277 of the Oil, Chemical, and Atomic Workers Union (OCAW), wasn't feeling too encouraged.

"I'm 39 years old and I have five kids. I've worked here 21 years, right out of high school. What do we know about the chemicals we work with, what they can do to us? There have always been fumes in the plant. We never wore masks. Sure, guys are thinking about quitting, but if you quit, you go to another plant, with other hazards. I just don't know."

— from *Medical World News*, May 3.

. . . The Labor Department promises it will promulgate a permanent standard [for vinyl chloride] at zero ppm by June. Even if a permanent standard is promulgated at zero exposure, it could be as ineffective as the standards regulating 14 occupational carcinogens set earlier this year—if, that is, its standards lack any monitoring provisions or allowance for medical surveillance of the workers.

— Barbara Newman in *The New Republic*, May 4.

. . . 94 consumer products containing vinyl chloride have been either voluntarily withdrawn by manufacturers or banned from sale by the EPA and FDA. These included 41 aerosol pesticides for household use, 26 home and professional hair-sprays, 13 deodorants and anti-perspirants.

— John Saar in *The Washington Post*, May 5.

At least three employes have been hospitalized and perhaps a dozen others have become ill in the last year and a half at a Milwaukee chemical firm apparently from inhaling chemicals. . . .

The two chemicals suspected of causing the problems are tetrachlora phalic anhydrite (TCPA) and benzophene tetracarboxylic dianhydrid (BTDA), used by the plant in the manufacture of plastic insulation or coverings for electrical components.

Officials of the Occupational Safety and Health Administration (OSHA) office here said they could find no mention of either of the two substances in any OSHA manuals or regulations.

So little is known about the substances, OSHA said, that no exposure limits have been set for workers.

— Neil D. Rosenberg and Joel McNally in *The Milwaukee Journal*, May 5.

Moving with unaccustomed speed under great pressure from certain segments of labor and Congress, OSHA has proposed a permanent standard of "no detectable level" for worker exposure to vinyl chloride.

Detection of any amount of vinyl chloride in the workplace would trigger work practices to prevent employee exposure, OSHA explained. This would apply to all manufacturing operations in which vinyl chloride is used. . . .

The Society of the Plastics Industry, which would be greatly impacted, immediately called the proposed standard "excessively and unrealistically restrictive" and said it will urge OSHA to hold hearings as well as to grant additional time to keep plants operating. A zero tolerance level would make it impossible for industry to operate, it said. It has commissioned Arthur D. Little, Inc., to prepare an economic impact survey that is expected to be completed in six to eight weeks.

— Gershon W. Fishbein in the *Occupational Health & Safety Letter*, May 8.

FLEMINGTON, N.J., May 8 — The Republic Steel Corporation expects 1974 to be a good year if the Reserve Mining Company's taconite plant at Silver Bay, Minn., is allowed to operate, William J. De Lancey, president, told about 50 stockholders at the annual meeting here today. . . .

Mr. De Lancey said that Reserve Mining, Armco and Republic are "convinced that no health hazard exists" at Silver Bay. . . .

The meeting lasted just 30 minutes. There were no questions from stockholders.

— Gene Smith in *The New York Times*, May 9.

The Senate defeated [tabled], 47 to 41, an effort to weaken the Occupational Health and Safety Act.

The amendment to an unrelated bill was sponsored by Sen. Peter H. Dominick (R-Colo.) and 29 others. It would have made penalties for serious employer violations permissive rather than mandatory, and would have stripped the Labor Department of power to impose fines for "non-serious" violations.

— from *The Washington Post*, May 9.

"Multi-corporate studies of cancer from vinyl chloride were initiated secretly in this country and in Europe. They were discussed with the government in secret and their results would probably still be a secret without the presence of mass media and if the law did not require the reporting of occupational disease. Even after the Louisville cases were announced, the government attempted to conduct key meetings without the presence of our representatives. In at least one case, in the Union Carbide plant in South Charleston, West Virginia, the government sought to substitute a medical survey of employees by an independent researcher, initiated by the Machinists union, with a study of their own to be conducted—by secret prearrangement—in collaboration with the company."

— Jacob Clayman, secretary-treasurer, AFL-CIO Industrial Union Department, speaking at a dinner meeting of the International Working Group on Vinyl Chloride, Delmonico Hotel, New York City, May 10.

Workers exposed to vinyl chloride, a widely used chemical in the plastics industry, face an increased risk of developing blood, liver and respiratory abnormalities as well as dying from a variety of cancers, the first international meeting on the chemical was told yesterday. . . .

A report from Romania indicated that nearly half of the vinyl chloride workers may have spasms in their blood vessels, and 10 per cent of the Romanian workers had temporary hormonal disturbances that in some cases were manifested as sexual impotence that disappeared during vacations.

The results of an intensive examination of nearly 400 workers at a Goodyear plant in Niagara Falls, N.Y., tends to support the European findings, according to data presented by Dr. Ruth Lilis of Mount Sinai Medical Center here.

Of 348 Niagara Falls workers whose lung function was studied, 58 per cent showed significant obstruction of the airways, Dr. Albert Miller of the Mount Sinai team reported. He added that this obstruction was found in a large percentage of nonsmokers as well as smokers, "suggesting that occupational exposure is a factor."

—Jane E. Brody in *The New York Times*, May 11.

NEW YORK — Scientists from around the world at a hastily organized meeting on the health hazards of vinyl chloride heard disturbing indications that the danger to workers in the plastics industry may be more widespread than first realized, and that the chemical may also pose environmental and public-health problems. . . .

. . . the most unsettling report came from scientists at Bonn University, who reported finding serious liver disease in six workers at a plant that turns PVC plastic into floor tiles. Until now, it had been thought the danger was limited to workers in direct contact with the vinyl-chloride monomer, who number about 6,500 in the U.S. But if the hazard extends to workers down the line who melt, mold, form, extrude, shape and otherwise fabricate PVC into pipes, packaging film and bottles, flooring, apparel, automotive parts, home furnishings, wire coatings and myriad other products, a much larger number of workers—some estimates say as many as 700,000—could be exposed to the danger. . . .

While several studies showed that the danger of vinyl chloride increased the longer the exposure to it, it isn't known whether low levels of exposure are safe. The Italian scientist, Dr. Cesare Maltoni, whose studies showed as early as 1972 that vinyl chloride at high levels in the air caused angiosarcoma in mice, but whose studies were ignored until the human cases were uncovered, has since reported that mice exposed to as little as 50 parts per million in the air also got the cancer. Fifty parts per million is the current emergency standard in vinyl-chloride plants set recently by the Labor Department, although the agency proposed last week tougher permanent standards that would limit air levels of vinyl chloride in plants to under one part per million. . . .

The EPA is currently surveying emissions of vinyl-chloride gas into the atmosphere and water found around factories nationwide to help determine if the health of people living around the plants may be endangered. Glenn E. Schweitzer, director of the EPA's Office of Toxic Substances, told the conference an estimated 200 million pounds of vinyl chloride are discharged into the atmosphere from U.S. vinyl-chloride-polymerization plants each year.

He called on industry to help support studies of populations around such plants who may have been exposed to low levels of vinyl chloride for long periods of time.

— Barry Kramer in *The Wall Street Journal*, May 13.

"Your Honor, I concede that the burden was on the Government in this to show that there was asbestos, to show that it's a carcinogen, and to show that it's in the water. At that point I seriously wonder if the burden doesn't shift. When you put a carcinogenic agent in drinking water and you show that some company is putting it there I say, I think it should be seriously considered that that burden shifts at that point.

"You poison a man's well, do you have to show dead bodies and so forth, or do you just say that you poisoned his well, it should be stopped."

— John P. Hills, attorney for the Government, arguing against the appeal of the Reserve Mining Company before the United States Court of Appeals for the Eighth Circuit, St. Louis, Missouri, May 15.

For at least a year, chemical firms in the United States and Europe withheld significant scientific findings linking liver cancer to gas used to make one of the commonest plastics, the

American Chemical Society's weekly news magazine reports today.

The gas is vinyl chloride, used to make the plastic polyvinyl chloride.

The trade association of the American chemical industry, the Manufacturing Chemists Association, joined in holding the findings in confidence, Chemical and Engineering News reports in its May 20 issue.

The Italian scientist who made the findings said that although preliminary, they were "predictive" of the fatal cancers recently recognized in several workers employed in converting the gas into polyvinyl chloride.

The scientist is Prof. Cesare Maltoni, director of the Istituto di Oncologia in Bologna. In a study initiated in September, 1971, by one of the European firms, Montedison, he discovered a rare form of cancer, angiosarcoma, in the livers of rats breathing vinyl chloride in a ratio of 250 parts per million (ppm).

That was in August 1972. At that time, the permissible level for workers allowed by the Occupational Safety and Health Administration (OSHA) was 500 ppm, or twice as much.

In January 1973, five months after Maltoni found the first rat angiosarcoma, a team of three U.S. chemical industry scientists visited him in Bologna to learn about his work.

The visit was preceded by months of negotiations with the European firms in addition to Montedison that had joined in sponsoring his studies: Imperial Chemical Industries, Solvay and Rhone-Progil.

But "U.S. chemical industry sources say they were bound by an agreement under which (the) four European firms controlled any release of animal test data obtained by Prof. Maltoni," Chemical and Engineering News says. It did not name any of the American firms.

— Morton Mintz in *The Washington Post*, May 20.

JUNE

Four new cases of angiosarcoma of the liver have just been uncovered. They raise for the first time the possibility in the minds of public health officials that this fatal cancer hazard that has been linked to vinyl chloride may extend to hundreds of thousands of workers and to members of the general public. . . .

Among the newly revealed cases of angiosarcoma of the liver—three in New York State and one in Connecticut—are a worker who for 30 years made electrical wire insulation from polyvinyl chloride resin and a woman who for nearly 30 years lived four blocks downwind from a polyvinyl chloride manufacturing plant.

— Jane E. Brody in *The New York Times*, June 1.

ST. LOUIS, June 4 — A three-judge Federal appeals court panel handed down today an opinion that apparently will allow the Reserve Mining Company to continue to pollute Lake Superior from its Silver Bay, Minn., plant for up to five years. . . .

The appeals panel said . . .

"We believe that Judge Lord carried his analysis one step beyond the evidence. Judge Lord apparently took the position that all uncertainties should be resolved in favor of health safety. The District Court's determination to resolve all doubts in favor of health safety represents a legislative policy judgment, not a judicial one. . . .

"The discharges may or may not result in detrimental health effects, but, for the present, that is simply unknown," the opinion said.

The judges concluded nevertheless, that "it must now be painfully clear to all who participated in the original decision to permit the discharge of tailings into Lake Superior that such a decision amounted to a monumental environmental mistake.

"The pollution of Lake Superior must cease as quickly as feasible under the circumstances," the court said.

— from *The New York Times*, June 5.

". . . I submit that if a million people in the so-called middle or professional class were dying each decade of preventable occupational disease, and if nearly four million were being disabled, there would long ago have been such a hue and cry for remedial action that if the Congress had not heeded it vast numbers of its members would have been turned out of office."

— The author, testifying before the House of Representatives Select Subcommittee on Labor, Washington, D.C., May 22.